SOTHIS PRESS

Sothis Press aims to completely reimagine the public domain book with an emphasis on universal access to captivatingly designed print editions. We believe that technology will enable every single book ever published to be perpetually accessible in beautiful print form. Sothis is beauty in ubiquity.

www.sothis.press
1 Waterhouse Square, London, ECIN 2ST

Machine Guns

Part I. Mechanism; Part II. The Practical Handling of Machine Gun Fire; Part III. Machine Gun Tactics

Julian S. Hatcher

SOTHIS PRESS

ISBN 978-1-0365-0155-6

This edition is published by
Sothis Press © 2023. The company
is registered in England and Wales.
The Company Number is 14900389.

MACHINE GUNS

PART I

MECHANISM

By CAPTAIN JULIAN S. HATCHER,

Ordnance Department, U. S. A.

PART II

THE PRACTICAL HANDLING OF MACHINE GUN FIRE

By 1ST LIEUT. GLENN P. WILHELM,

4th Infantry

PART III

MACHINE GUN TACTICS

By 1ST LIEUT. HARRY J. MALONY,

26th Infantry

The Collegiate Press

GEORGE BANTA PUBLISHING COMPANY

MENASHA, WISCONSIN

INTRODUCTION

In the summer of 1916, the War Department gave authority for the establishment of several Machine Gun Schools of Instruction for troops serving in the Southern Department. The first school to begin operations under this authority was established by Lieut.-Colonel C. C. Williams, Ordnance Department, at Harlingen, Texas. Two of the present authors, Capt. J. S. Hatcher, Ordnance Department, who had previously been engaged in the preliminary machine gun instruction of troops at various points along the border, and 1st Lieut. H. J. Malony, 26th Infantry, who had served for several years in command of various machine gun organizations, were assigned as instructors. These officers were later joined by 1st Lieut. Glenn P. Wilhelm, 4th Infantry, who was in command of the machine gun company of that organization, and 1st Lieut. W. W. Doe, 26th Infantry. At the present time the school has been in operation for over six months, and the experience of the authors in the operation of the guns at the school has covered an average expenditure of about 10,000 rounds of ammunition per week during this period.

The notes made by the authors during their work at the machine gun school and elsewhere form the basis for this volume. These notes were rewritten and arranged in their present form by Captain Hatcher.

It is desired to give credit to Lieutenant Doe for much valuable advice and assistance.

Not all of the material contained in this volume is original, as much which was compiled from various sources is collected here to afford an opportunity for ready reference.

It should be understood that this book represents only the personal views of the authors, and is in no degree official.

J. S. Hatcher.
G. P. Wilhelm.
H. J. Malony.

Machine Gun School,
Harlingen, Texas,
March 15, 1917.

PART I

NOTES ON THE MECHANISM AND MECHANICAL HANDLING OF MACHINE GUNS

Before proceeding to a discussion of the mechanism of the various types of machine gun, it is desired to speak of the following points which are equally applicable to the various types of machine guns treated in this book:

Jams, Malfunctions, Stoppages

Distinguish carefully between these terms, and use them correctly. Any accidental cessation of fire is a *stoppage*. It may be due to a misfire, or to the fact that the magazine has been emptied, etc. In this case it is not a malfunction.

A *malfunction* is an improper action of some part of the gun, resulting in a stoppage. For example, a failure to extract the empty cartridge case.

A *jam* is some malfunction which causes the mechanism to stick or bind so that it is difficult to move. Do not use the word "Jam" too much. Most troubles with the guns are merely temporary stoppages due to some malfunction, and real jams are comparatively rare.

"*Headspace.*" This is a term frequently used by machine gunners to denote the malfunction caused by the fact that the space for the head of the cartridge, between the rear end of the chamber and the front of the breech-block, is too great. This causes the cartridge to rupture, and only the rear end is extracted, the front end being left in the chamber. The reason that excessive headspace causes ruptured shells is because, when the explosion takes place, the powder pressure causes the forward end of the cartridge case, which is thin, to expand and grip the chamber, and if the rear end is not firmly supported by the breech-block, it is blown off. See the accompanying sketch.

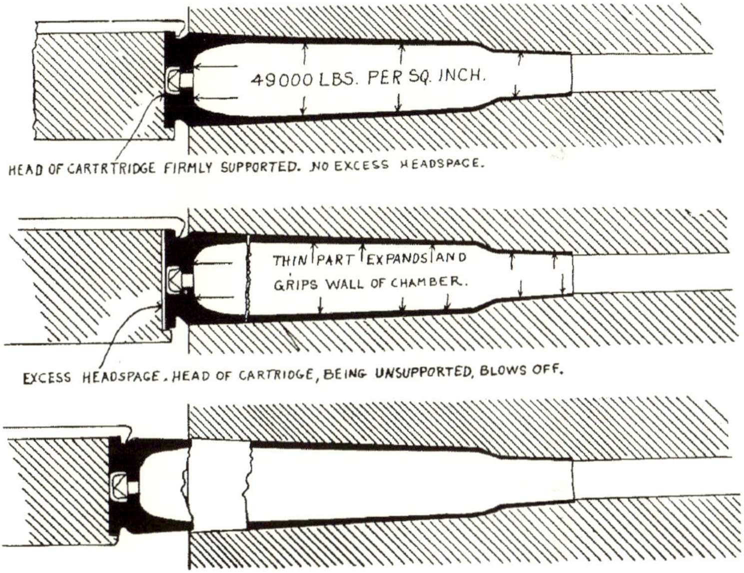

THE RESULT OF EXCESS HEADSPACE.

The top figure shows a cartridge properly supported in the chamber. The second figure shows a cartridge in a gun with excess headspace. The third figure shows the same cartridge being extracted after the head has been blown off, showing part of the shell left in the chamber, to interfere with the seating of the next cartridge.

ORDNANCE DEPARTMENT HANDBOOKS

Much trouble will be avoided if the Ordnance Department pamphlets for the various guns are carefully studied and the directions given in them are followed.

CHAPTER I

THE BENET-MERCIÉ AUTOMATIC MACHINE RIFLE

This weapon, which has been in use in the United States Army since 1909, was the first automatic machine gun sufficiently light in weight to be portable and at the same time sufficiently durable to be serviceable.

The gun is of the gas-operated, air-cooled type, and when in competent hands, is exceedingly reliable. Aside from its durability, its strongest point is the case with which ammunition can be carried packed in the clips (called feed strips) ready for shooting. Another strong point is the ease with which barrels may be changed without disturbing the rest of the mechanism.

The photograph shows the appearance of the gun. For the names of the parts, see the sectional view.

The operation of the mechanism in firing is as follows:

When the rifle is fired and the bullet has passed the gas port in the barrel a part of the powder gas passes into the chamber of the gas-nozzle ring, and enters the gas cup of the actuator, forcing the latter to the rear. The actuator in recoiling compresses the actuator spring, the cam surface cut in its upper portion engaging the lug of the fermeture nut, causing the latter to rotate so as to disengage its threads from those of the breech-block. At this part of the recoil the firing pin, which has been drawn back by the actuator, engages its upper lug in the ramp of the receiver. The firing pin then rotates on its axis and its upper lug comes to rest in the transverse cut of the breech-block, thus locking the latter to the actuator. The large lug of the

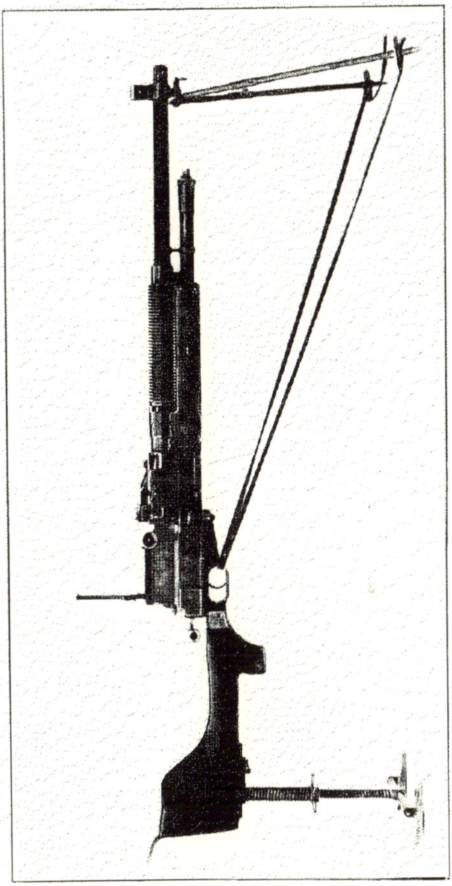

actuator strikes against the shoulder of the breech-block, drawing the latter to the rear and thereby completing the opening of the breech. The claw of the extractor engages the groove of the cartridge case and draws it from the chamber. During the recoil of the breech-block the head of the cartridge case strikes against the ejector, throwing the case out of the rifle through the ejection opening in the receiver. When the actuator is partly recoiled the cam surface cut on its right side engages the upper lug of the feed piece, causing the latter to rotate from right to left on its axis. The feed arm of the feed piece engages its feed lug in the central opening of the feed strip, forcing the latter into the rifle and placing a cartridge in the loading position in front of the chamber. The pawl of the feed-piece spring engages in the lateral openings of the feed strip, thereby holding the latter in place and preventing its being drawn to the right when the feed arm of the feed piece, having advanced the feed strip one notch, returns to its original position by sliding over the feed strip in order to engage in the next hole. Finally, when the actuator is almost recoiled (supposing the rifle to be set for single shots), the sear engages in the cocking notch of the actuator and the latter is held back ready for the next shot.

When the trigger is pulled the sear is disengaged from the actuator, which, now being free, is thrown forward, due to the force exerted by the actuator spring. During this movement the actuator carries forward the breech-block, to which it is connected by means of the large lug of the firing pin. The breech-block strikes the cartridge, forces it into the chamber, and the claw of the extractor engages the groove of the cartridge case. As soon as the breech-block engages the fermeture nut the large lug of the firing pin engages in the ramp of the receiver, causing the firing pin to rotate so as to disengage this lug from the transverse cut of the breech-block. The firing pin is then free and may move forward. The actuator continues its forward movement and its cam face engages the lug of the fermeture nut, rotating the latter so as to cause its threads to engage with those of the breech-block. The breech is now closed and locked. The firing pin striking the primer fires the piece. During this forward movement of the actuator, the small cam surface cut on its right side engages the lower lug of the feed piece, causing it to rotate from left to right. The feed arm is actuated by this movement, and its pawl glides over the feed strip and engages in the next opening, ready to feed the strip another notch into the rifle when the actuator again recoils.

A detailed description of each part of this rifle is given in Ordnance Pamphlet No. 1926. A short discussion of the common difficulties likely to be experienced with each part is given below.

Barrel—The orifice may become stopped up. A partially stopped orifice will cause the rifle to take excessive gas, and one that is completely stopped will cause the rifle to fail to function at zero gas setting. To remedy the difficulty, remove the orifice screw and clear the vent with an .087″ drill or reamer.

If the gas chamber of the gas-nozzle ring has been reamed excessively with the gas cylinder cleaner, the gun may give symptoms of a partially stopped vent. The remedy is to obtain a new barrel. Care should, of course, be taken to prevent this abuse of the rifle. This can be accomplished by forbidding the use of the gas cylinder cleaner except when necessary, and by instructing the personnel of the company as to the effects of the careless use of this tool.

The sharp edges of the barrel at the entrance may shave chips of brass from the cartridges and so clog the mechanism.

These sharp edges should be dulled with a small scraper or a half-round file.

The chamber may become enlarged through wear or corrosion. This will allow the cartridge to go too far forward, and will give excessive headspace. The remedy is to obtain a new barrel.

Locking Nut—The locking nut may become loose through wear or from the stress of firing. This will allow the barrel to go forward slightly and cause excessive headspace. The remedy is to obtain a new locking nut. For an emergency remedy the locking screw should be removed and the locking nut turned as far past the locked position as possible. This will draw the barrel more tightly to the receiver. The locking screw should then be replaced and screwed up as tightly as possible. As the locking nut has been screwed past its normal position, the hole in the nut will no longer register with the one in the receiver, and the end of the locking screw cannot enter the hole in the receiver, but will strike the surface of the receiver below the hole. If it is screwed up as tightly as possible it will bind against the receiver hard enough to keep the locking nut from unscrewing. If the gun is to be used for some time in this condition, the locking screw will hold better in its new position if it is cupped slightly in the end with a prick punch.

The old model locking nut can not be tightened as indicated above, as the stop surfaces are so placed as to prevent the nut from being turned past the normal position. Locking nuts should be watched carefully and replaced at the first indication of looseness, as even if the headspace is not sufficient to rupture shells it will still cause a pounding effect which will strain every part of the gun. Guns have been found in which the pounding from this cause had been allowed to continue until the barrel, locking nut, fermeture nut, breech-block, and receiver had been strained so badly that each part was unserviceable.

Another trouble that is often caused by shooting the gun with a loose locking nut is the battering of the cup on the actuator, due to the fact that the looseness of the locking nut allows the gun to sag in the middle so that the actuator strikes the gas nozzle when the gun is fired. This will also cause the gas nozzle to become battered. In guns that have been subjected to this abuse, the gas will escape from the actuator cup without exerting its full force, so that an excessive amount of gas will have to be used to blow back the actuator. The gun will give symptoms similar to those caused by a stopped orifice, and may fail to operate with zero gas setting.

Fermeture Nut—This article rarely gives trouble, provided that the new model nuts are used. When breakages do occur, they are usually due to the pounding effect of excessive headspace. The new model fermeture nuts are marked "1914" on the bottom lug.

Actuator—The new model actuator is marked "1914" on the rear end. If a gun is fired with a loose locking nut the actuator will strike the gas-nozzle, and may become bent, or have the cup on the end enlarged. An actuator which has been subjected to this abuse will require more gas to drive it back, and the gun will give symptoms of a partially clogged vent. In case of doubt, examine the inside of the cup on the end of the actuator and see if it has been striking the gas-nozzle.

Actuator Spring—After continued use the actuator spring may take a permanent set and become too weak. This may be detected by comparing it with a new spring. A spring which has become weakened should be replaced. It is never necessary to weaken the actuator spring. Cases have been known where, when guns would not operate at zero gas setting, the actuator spring has been clipped. This is wrong. If the gun will not operate at zero gas setting with a normal actuator spring, there is some trouble, such as a clogged orifice or a bent actuator, and the trouble should be located and removed.

Hand-guard—This part sometimes gives trouble through being wrongly assembled. The front stiffening piece gets

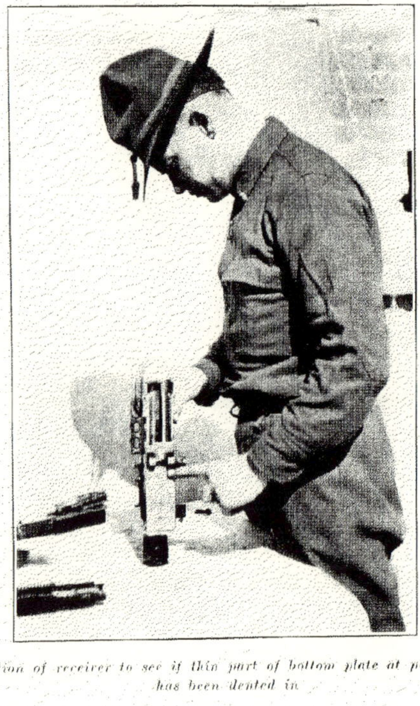

Examination of receiver to see if thin part of bottom plate at point indicated has been dented in

below instead of above the lug on the gas-nozzle ring, and the front end of the hand-guard is held low enough to interfere with the actuator. The result is that while the gun can be cocked and snapped, no explosion of the cartridge occurs, as the actuator is stopped before the end of its movement and the firing pin is prevented from striking the cartridge. This gives symptoms similar to a broken firing-pin point.

Receiver—Between the lugs on the bottom of the receiver there is a thin spot which sometimes gets dented in so as to interfere with the actuator. The cause of this accident is usually carelessness in disassembling the guard group from the receiver. If a receiver is found to be in this condition, the remedy is to smooth out the dent from the inside by the use of any suitable metallic instrument. The photograph shows the location of the dent.

Guard Group—Trouble is sometimes caused by the loosening of the guard screw and stock bolt. This allows the stock to become bent up slightly and so interfere with the cocking handle disc in cocking the gun. These screws should be tightened periodically.

Examination of the Gun Before Firing—If the gun is carefully examined periodically and noticeable faults are corrected, malfunctions on the firing line will be reduced to a minimum. The examination is best conducted as follows:

Cock the gun, and then let the mechanism down slowly by holding on to the cocking handle. If, when let down slowly in this manner, the mechanism shows a tendency to stick at any point, it shows that there is undue friction at that point. The cause should be discovered and removed.

Take out the locking screw and cocking handle, and remove the guard group from receiver. Test out the trigger mechanism as follows:

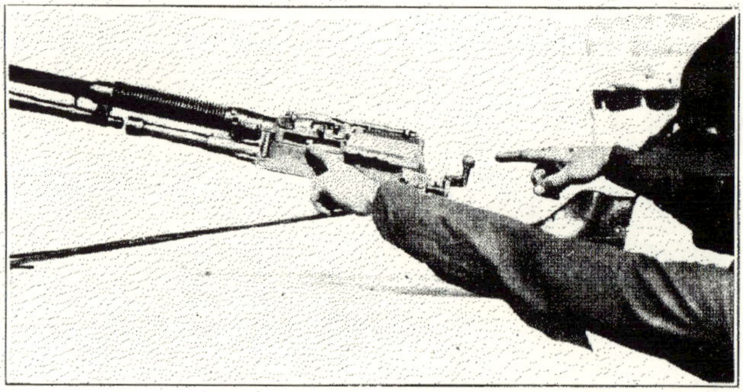

Undue friction in the mechanism. Note the position in which fermeture nut and cocking handle stopped when the mechanism was let down slowly.

Testing trigger mechanism

With trigger and sear in place, put the cocking handle in the guard, and turn it to "R." Hold the cocking handle disc firmly against guard with both thumbs, and pull the trigger. The trigger arm should pull back the sear and then release it. Holding the trigger back and holding the cocking handle disc firmly against the guard, take hold of the sear knob and turn it to see how much the sear arm clears the hook on the trigger arm. It should just barely clear. If the parts clear by more than the thickness of one or two sheets of paper, it shows that the trigger arm is bent up. If the trigger pulls the sear back and fails to release it with the cocking handle disc set at "R" it shows that the trigger arm is bent down. Either fault may be corrected by bending the trigger arm up or down as the case may be. *Never attempt to correct a defect of this nature by filing the trigger; it will only make matters worse.*

Take out the actuator, breech-block, and firing pin.

Examine the actuator for burrs. If any are found, correct them by pounding the burr down carefully with a machinist's hammer, afterwards smoothing the surface with emery cloth or a file. The use of a file on the working surfaces of the actuator should be avoided, as these surfaces are case hardened, and filing will expose the soft metal under the surface.

Examine the breech-block to see if it holds cartridges tightly during extraction. This test is made by placing a live cartridge so that it is held by the extractor against the face of the breech-block in a position similar to that occupied at the moment of firing. When the cartridge is properly held in its seat in the breech-block it will not fall out when the block is shaken slightly, no matter in what position the block is held. If the breech-block fails to hold the cartridges tightly, it will sometimes drop empty shells in the receiver and cause jams. If the cartridge is loosely held, the fault may lie with the extractor spring. A weak or broken spring will cause these symptoms. A worn extractor will also cause

Two points at which burrs on actuator are likely to be found

this trouble. When this trouble is encountered the extractor spring should be changed and the test repeated. If the trouble persists, the extractor should be changed. If this does not remedy the trouble, it is due to the breech-block.

Wear on the front lips of the breech-block on each side of the ejector groove will cause this difficulty. If trouble is found to be in the breech-block it can often be corrected by slightly

Testing to see if breech-block holds cartridge firmly

upsetting the front lips of the breech-block on each side of the ejector groove with a machinist's hammer.

Test firing pin for length by inserting it in breech-block to see if it projects through far enough to fire a primer. Frequently the point of the firing pin will break with a conical fracture such that the breakage will not be evident to the naked eye, while the firing pin may be too short to fire a primer with certainty.

Remove the feed piece spring. Examine the feed piece spring pawl to see that it is tight. Examine the vertical face of the feed piece spring pawl to see if it has been notched by the feed strips. If so, the notch should be removed by filing as otherwise it may catch on the feed strip.

Take out the feed piece, remove barrel, hand guard, locking nut, and fermeture nut.

Examine fermeture nut for burrs on the teeth or on the bottom lug. If any are found they should be removed. Examine the wing of the fermeture nut for a crack. If one is discovered, use a new nut.

Examine receiver inside to see that the thin bottom surface just forward of the lugs for guard trunnions is not bent in so as to interfere with actuator. If it is, the dent should be smoothed out with any suitable implement.

Enter a live cartridge under the stripping finger of the receiver (called "tongue" in handbook) to see that it is not bent up or down. A live cartridge should enter with enough extra space between it and the stripping finger to allow the insertion of a penny. Any defect can be corrected by bending the tongue up or down. In doing this, great care must be used, as the receiver cannot be repaired if this part is broken.

Examine rear sight to see if it is loose on the receiver. If so, tighten it by screwing up the rear sight fixed base screw. After it has been tightened, bind the screw in place by upsetting the edge slightly with a center punch.

Examine the edges of the barrel at the entrance to the chamber, and if they are sharp, dull them with a half round file or a scraper, so that they will not shave strips of brass off of the entering cartridges.

Tighten the guard screw and stock bolt if possible. If these parts are neglected, the stock will become bent up so as to interfere with the cocking handle disc in retracting the actuator.

Examine the elevating mechanism to see that the binder is properly adjusted. The adjustment is made by turning the binder lever pivot with a screw driver. The part of the binder lever pivot on which the binder lever works is eccentric to the part of the pivot which fits in the elevating screw slide, and the high point of the cam is indicated by an arrow stamped on the head of the pivot. To tighten the binder, turn the pivot so as to bring the point of the arrow toward the rear of the gun. To loosen the binder turn the pivot so as to bring the point of arrow toward the muzzle of the gun. The binder lever pivot should be a driving pit in the elevating screw slide, and if it shows signs of looseness, it should be replaced.

Replace the fermeture nut and locking nut, and assemble the barrel to the receiver, leaving off the hand guard. In screwing up the locking nut after replacing the barrel, see if it can be turned past the locked position by hand. If it can, the locking nut is too loose and should be replaced. Turn the locking nut to the locked position and put in the locking screw. Lay the barrel of the gun on a table or bench with the receiver projecting over the edge, and let one man hold the barrel down firmly while another shakes the receiver to see if any play can be detected between the barrel and receiver. Any looseness at this point is due either to a loose locking nut, a worn barrel, or a worn receiver, and should be corrected before the gun is fired, as otherwise the rifle will be badly strained from the pounding effect of exces-

Testing to see if locking nut is tight

sive headspace. A temporary method of tightening this joint is to turn the locking nut past the locked position as far as it will go. In this position the locking screw will not go into the hole for it in the receiver, but it will bind against the receiver and if screwed up tightly will hold the nut from jarring loose.

Assemble the rifle except the handguard. Support the piece on the barrel rest and the elevating mechanism as for firing, and move the actuator backward and forward with the cocking handle to see if it strikes the gas-nozzle. A bent actuator will strike against the nozzle, and the motion will be arrested. A loose locking nut will also cause the actuator to strike the gas-nozzle.

Put on the handguard and oil the mechanism well and the rifle is ready to go to the firing point.

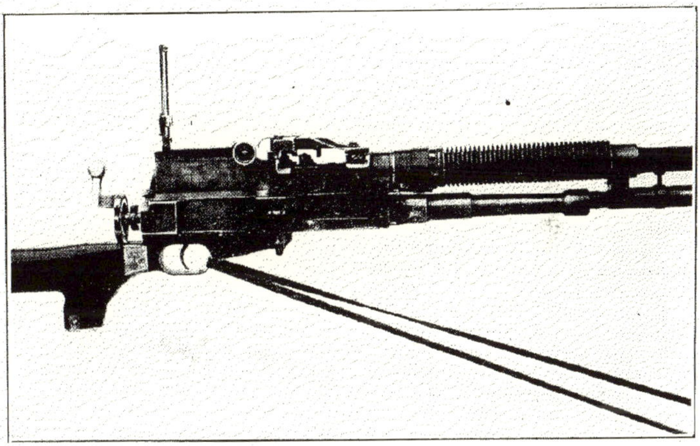

Showing how gas cup of actuator will strike gas nozzle if loose locking nut is used

Determination of Gas Regulator Setting—The determination of the proper setting of the gas regulator is most important, as either too much or too little gas will cause stoppages of fire. One symptom of too little gas is the failure of the actuator to return far enough to throw out the empty shell. The gun stops and the breech-block is found to be about half way forward with an empty shell between it and the back of the barrel. If more gas is used, but still not enough to operate the gun properly, the actuator may return far enough to eject the empty shell and feed the live one without having the actuator blown back far enough to catch on the sear. This will cause the gun to fire more than one shot while the cocking handle is set at "R."

To determine the gas setting, proceed as follows:

Place the rifle on the firing point, cock it, and enter a strip of cartridges. Remove the regulator entirely and fire one of two shots to clear the orifice. It will of course be necessary to cock the rifle by hand after each one of these shots. Put in the regulator and set it at 4.0. Set the cocking handle at "R" and fire one shot. If the gun shows symptoms of not enough gas, screw down the gas regulator several turns and try again. Repeat this until the gun shows no further symptoms of too little gas. Then screw the regulator down two turns to provide enough gas to insure the operation of the gun when the oil begins to gum from the heat or when sand or dust gets into the gun. This setting should be carefully recorded and the gas regulator should be habitually left at this reading.

Too much gas will also cause trouble. If the regulator is screwed down too far it will cause the gun to race, and sometimes the gun will stop with two cartridges, a live one and an empty one, jammed in the receiver.

This is usually due to one of the following causes:

(a) The speed is so great that the extractor tears the rim partly off of the cartridge instead of extracting it, or

(b) The ejector cuts through the rim of the cartridge instead of ejecting it, or

(c) The feed strip is fed into the gun so fast that the downward impulse given to the cartridge by the stripping finger is sufficient to drive it down into the receiver on top of the cartridge about to be ejected.

Marks on the head of the cartridge where the ejector has struck are a sign of too much gas.

A sharp dent on the cartridge case near the shoulder where the case strikes the telescopic sight bracket in ejection is not necessarily a sign of too much gas.

The sizing of Feed Strips—When the feed strips have been in use some time the clips that hold the cartridges become set so as to hold the cartridges insecurely. If strips in this condition are fired, malfunctions are likely to occur. The gun will stop and it will be found that two cartridges are jammed into the receiver. This is caused by the fact that the cartridges are held so loosely in the strip that the vibration of firing causes them to jar down into the receiver before the empty shell is ejected. To obviate this difficulty the strips should be sized down after each time they are fired.

Only after considerable experience can the proper degree of tightness of the strips be recognized by pushing cartridges out of the clips with the fingers, and usually reliance will have to be placed on some method of gauging the setting of the

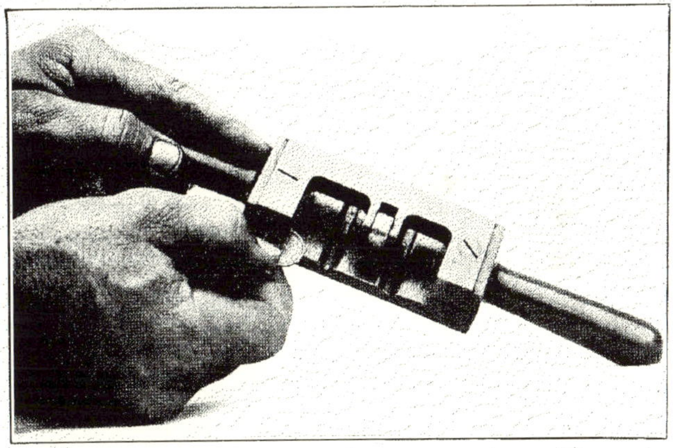

resizing tool to insure the proper resizing of the strips. A very satisfactory method of gauging the setting of the resizing tool is to adjust it so that a piece of metal of the thickness of a penny (0.05 inch) can just be inserted between the support and the body of the tool. The method of gauging this is shown in the photograph.

As the middle clip of the feed strip is not in the center, it is possible to insert the strip in the reloading tool so that the roller passes over one side of the middle clip. This will not resize the strip, and care should be taken to enter the strip properly in the tool.

TELESCOPIC SIGHTS:

This instrument, if properly used, is a great aid to the effectiveness of the Benet Gun, as by its use the line of sight is removed from above the hot barrel and the heat mirage is avoided. The telescopic sight is also an aid to target recognition, and very often allows the gunner to observe the strike of his own shots.

On the reticule glass is a cross for indicating the line of sight and three stadia lines which are intended as an aid to the estimation of distances, as the angle included between the horizontal crosswire and each of the three lines subtends the average height of a man (68 inches) at the distance of 1000 yards, 1500 yards, and 2000 yards, respectively for the three lines.

Adjustment of the Telescopic Sight:

For focus—In adjusting the instrument at the factory the reticule is set exactly in the focal plane of the objective, which has a universal focus beyond 100 feet, and the eyepiece is set for the vision of an average observer. Should alteration of the focus of the eyepiece be necessary to suit special conditions, unscrew the focusing lock nut and screw the eyepiece out or in until the cross lines of the reticule are sharply defined, and at the same time the image of the target is clearly visible. Then move the head up and down, so that the eye may travel across the eyepiece. If the focus has been correctly found there will be no parallax—that is, no apparent motion of the cross wires with reference to the target when the eye moves across the field. The focus being accurately adjusted, be sure to lock the eyelens holder by tightening the focusing lock nut. The rubber eye cap can now be turned to any desired position without affecting the focus. If the rubber cap does not turn easily, loosen the eye-cap ferrule and tighten it again when the adjustment is made.

For elevation.—The range dial is graduated from 0 to 3,000 yards by 20-yard divisions. To make elevation adjustment set the sight of the rifle at 500 yards. Set the range dial of telescopic sight at the same range and clamp it. The cross wires of the telescopic sight and the line of sight of the rifle through the open sight should then bisect the target exactly. If the cross wires of the telescopic sight are too high, unscrew the hexagon adjusting-screw nut, using the wrench provided with the sight, and turn the adjusting screw clock-wise, which will cause the cross wires to drop on the target. Turning the adjusting screw in the opposite direction will make the cross wires rise. After the adjustment is made be sure to tighten the nut. The rifle is then fired several shots to test elevation, the necessary correction being made by the movement of the adjusting screw until a satisfactory elevation is obtained.

For drift.—The graduations on the drift dial correspond to 1-inch on the target at a range of 100 yards. Turning the dial clockwise corrects to the right, as indicated by the letter "R," and counter clockwise to the left, as indicated by the letter "L." There are 38 points of left and 46 points of right drift correction. The drift dial should read "O" when the optical axis of the instrument is parallel with the bore of the rifle. Should it be necessary to adjust the dial, make the optical axis of the instrument parallel with the bore of the rifle, loosen the two drift-dial screws, which will permit rotating the dial without moving the drift screw, set the dial exactly at "O," and clamp the screws firmly again.

Care and Preservation: Telescopic sights are necessarily delicate instruments and must not be subjected to rough usage, jars, or strains. When not in use the telescopic sight should be kept in its case and stored in a dry place. It should be occasionally examined to insure its not being corroded, and all traces of dust or moisture should be removed before it is put away. To obtain satisfactory vision the glasses should be kept perfectly clean and dry. In case moisture collects on the glasses, place the telescope in a gentle warmth; this is usually sufficient to remove it. A piece of chamois skin or a clean linen handkerchief will answer for cleaning purposes, care being taken that the cleaning material does not contain any dirt or grit. These sights before issue are carefully adjusted and their adjustment proven by actual firings. The prism holder should never be opened except by a competent person. The body of this telescope and its objective must remain intact. The eyepiece can be removed after loosening the focusing lock nut and the eyelens and reticule then cleaned. Before incasing the prisms at the factory the interior of the body is thoroughly cleaned and all particles of dust removed. If any fine particles should be left in the body, or if the body should be opened and particles enter, they will settle upon the reticule and, when magnified by the eyelens, obscure the vision. Four small holes are punched through the rubber eye cap to permit the escape of air on recoil, and to prevent suction on counter recoil.

THE USE OF BLANK CARTRIDGES IN THE BENET GUN:

In Armories or other places where ball cartridges cannot be fired, valuable instruction in the operation of the mechanism can be obtained by using blank ammunition. This is also desirable in manoeuvers, sham battles, etc. As the Benet gun is at present issued, it will not operate with blank cartridges, but it can be made to operate with special blank

Attachment for firing blank cartridges

cartridges if a gas check is attached to the muzzle. The photographs show a gun with an attachment for this purpose which was improvised by the Machine Gun Troop, Squadron "A" New York Cavalry, under the supervision of Captain Henry Sheldon, Commanding Officer of the Troop.

The attachment is simply a cylinder screwed onto the muzzle of the gun, and forming an extension of the barrel. The bore of the cylinder tapers from .30 cal. at the rear end to .16 cal. at the front end, thus throttling the gas and increasing the pressure in the bore. Where the cylinder abuts against the end of the barrel a copper washer is inserted to prevent the escape of gas.

To give the best results with this attachment, the cartridges should have a heavier charge than that in the regular issue of blank cartridges. With the Benet gun 30 grains of Du Pont Ballistite has been found to give satisfactory results if used in blank cartridges fitted with the waxed paper bullet formerly used in the Army. With the Colt gun the best load is 17½ grains of Infallible powder.

While this attachment is not issued by the Government, probably because it will not operate with the service blank ammunition, it is mentioned here as a matter of interest and information.

COCKING THE PIECE:

The novice sometimes finds it hard to cock the gun, owing to the lack of instruction and practice in this work. When the proper method is employed, the gun can easily be cocked without necessitating any change in the gunner's position.

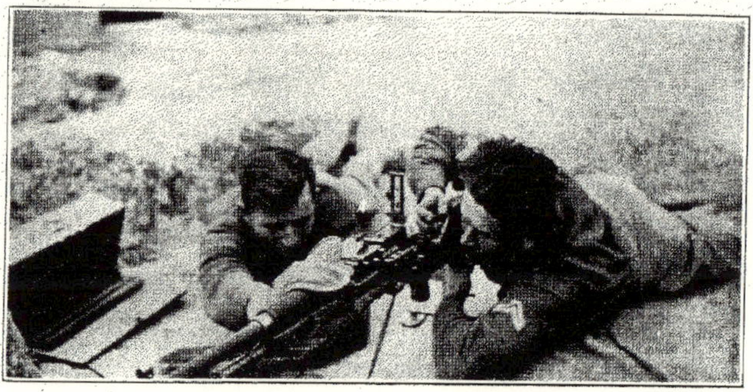

The proper method of cocking the piece

Accordingly all members of the gun squads should be practiced in cocking the gun while it is in the firing position on the ground until they can do so easily and quickly without moving the barrel rest or deranging the aim. The proper method, which is shown in the accompanying photograph, is as follows:

The loader steadies the piece by placing his right hand on top of the radiator. As the barrel may be hot, he uses an asbestos glove on this hand. The gunner grasps the cocking handle with his right hand and the outer elevating screw with his left hand. He turns slightly on his left side, and holding the gun forward with his left hand, he snaps back the cocking handle with a quick motion of his right hand, cocking the gun.

THE USE OF BOLOS TO STEADY THE BARREL REST:

When the Benet Gun is used on the field mount the front legs may be moved in cocking the gun, loading, or reducing

The method of using bolos to steady the barrel rest

a jam. This accident, which will cause a serious loss of time owing to the necessity of resetting the barrel rest and relaying the gun, can be prevented by driving a pair of bolos behind the barrel rest legs, as shown by the photograph.

LOADING:

In loading the gun, great care should be taken to get the feed strip properly entered. If the front edge of the feed strip should get under the front feed guide, a jam will result. The gun must be loaded as quickly as possible in order to avoid a loss of time between strings. Care must be taken in loading not to derange the gunner's aim. In order that the loader may accomplish the results mentioned above, he should be given abundant practice in loading the gun. In order to avoid a large expenditure of ammunition it is common for this practice to be carried out with strips that have been loaded with only three cartridges in each end. In this practice the strip is entered in the gun, the first three shots are fired, the gun is recocked, and the strip is pushed in until the first of the remaining three cartridges is in the firing position. The gunner fires these three shots, the loader enters a fresh strip, and the gunner fires the first three shots, all as quickly as possible. For this exercise, time should be taken with a stop watch. All shots should be fired at a target, and hits should be counted. This will prevent the sacrifice of accuracy to speed, and will teach the loader not to derange the gunner's aim.

The proper method of loading is shown in the photograph. The loader lifts the feed piece and places his left hand on the receiver, holding it firmly to prevent deranging the gunner's aim. He places his fingers on the cartridge stop, and his thumb against the rear feed guide. This thumb serves as a guide for entering the strip at night. With the right hand the strip is entered slightly, and the outer end of the strip is then pressed well down. The first cartridge acts as a ful-

The proper method of loading. Note that outer end of strip is pressed down to prevent front end of strip from getting under feed guide.

crum, and the leverage thus applied forces the entering edges of the feed strip up so that there is no chance of their getting under the guides. The strip is then forced in until the cartridge stop projects. If the feed piece does not drop, the loader forces it down with the first finger of his left hand. He removes his hand, and the gun is ready to fire.

COOLING THE GUN:

If water is available the barrel should be cooled after about 300 rounds have been fired. This is done by applying water to the radiator of the barrel with the cooling sponge. Another method of cooling, which is very effective when circumstances permit it, is to lift the gun up and dip the muzzle into a cup or bucket of water. The formation of steam will cause the water to geyser up through the barrel, cooling it rapidly. In case water is not available, the barrel should be changed after about 500 rounds, if practicable.

If it is impracticable to cool the gun or change barrels, 1000 consecutive rounds may be fired without permanently injuring the gun.

POINTS TO BE OBSERVED BEFORE GOING ON THE FIRING LINE FOR TARGET PRACTICE OR ACTION:

(a) Examine mechanism to verify readiness for action, as described above. Correct any defects discovered.

(b) See that binder is properly adjusted.

(c) See that working parts are well oiled, and that no oil is allowed to get on outside of barrel.

(d) See that bore and chamber are free from dirt or foreign substances.

(e) Set gas regulator to proper reading.

(f) See that latigo straps for barrel rest are properly adjusted.

(g) See that spare barrel is in good condition, bore clear, and gas regulator properly set.

(h) See that telescopic sights are clean and properly adjusted.

(i) See that oil cans are full.

(j) See that contents of gunner's pouch and spare parts case are in good condition.

(k) See that no worn or deformed parts are allowed to get into gunner's pouch or spare parts case.

(l) Be sure that feed strips are properly sized.

(m) See that ammunition is clean and free from grease or dirt.

(n) Have cooling sponges in good condition and ready for use.

POINTS TO BE ATTENDED TO DURING FIRING:

(a) Lay the gunner's pouch on the ground near the gun. If it is worn on the body it will be forgotten and left with the gunner if he should become a casualty.

(b) During a cessation of fire replace a partly used feed strip with a full one, and cool the gun or change the barrel if time permits.

(c) Embrace every opportunity to oil the working parts of the gun.

(d) Replenish the supply of ammunition on the firing line at every opportunity.

(e) During a temporary cessation of firing, reset the barrel rest legs if necessary.

(f) When barrel gets very hot, cool it with cooling sponge or change barrel.

(g) Watch the ejection to see if gas setting is all right. When ejection becomes weak, screw down the regulator several turns. This becomes necessary after a few minutes firing, owing to the increase of friction which occurs when the oil burns off or becomes mixed with blowing dust.

(h) If sand is blowing, cover up the receiver when the gun is not firing.

(i) If telescopic sight is being used, verify the setting from time to time, as the vibration of firing frequently causes the windage setting to change.

POINTS TO BE ATTENDED TO AFTER FIRING:

(a) Before leaving the firing point, remove the feed strip and snap the mechanism on an empty chamber to be sure the gun is unloaded.

(b) Clean and oil the bore of the gun immediately after firing and once a day thereafter for ten days.

(c) When time permits, dismount the gun, clean and oil all parts, correct any defects that may have been discovered during the firing, and reassemble.

(d) Tighten stock bolt and guard screw.

(e) Resize and reload all feed strips.

JAMS, MALFUNCTIONS, AND STOPPAGES:

The most important causes of malfunctions in the Benet gun are, in the order of their importance:

The use of loose locking nut.

Improper gas adjustment.

Improperly sized feed strips.

Fortunately, all of these causes are avoidable, and consequently, with a well instructed gun squad malfunctions are of rare occurrence.

While the causes mentioned above are responsible for perhaps ninety per cent of the stoppages, there are, of course, numerous things that can cause trouble, and as far as practicable these are shown in the following table:

I. *Improper Manipulation:*	(a) Wrong gas adjustment, not enough gas. (b) Wrong gas adjustment, too much gas. (c) Feed strip improperly entered in loading. (d) Feed strip too loosely sized. (e) Dirt or grease in the chamber. (f) Dirt or grease on cartridges. (g) Dirt in fermeture nut. (h) Ejector cap improperly assembled. (i) Handguard improperly assembled.
II. *Breakages:*	(a) Fermeture nut. (b) Locking nut. (c) Firing pin. (d) Extractor spring. (e) Breech-block (corner broken). (f) Ejector spring (rare). (g) Feed piece (rare).
III. *Deformation of parts caused by firing gun with loose locking nut:*	(a) Loose locking nut, causing ruptured shells. (b) Strained receiver, causing ruptured shells. (c) Strained fermeture nut, causing ruptured shells. (d) Strained breech-block, causing ruptured shells. (e) Broken fermeture nut. (f) Actuator cup worn from striking gas nozzle. (g) Gas nozzle worn from being struck by actuator.

IV. *Deformed or worn parts, other than those given under III :*	(a)	Bent handguard.
	(b)	Enlarged chamber.
	(c)	Worn breech-block (Drop shells).
	(d)	Worn extractor.
	(e)	Weak extractor springs.
	(f)	Stopped orifice.
	(g)	Worn firing pin.
	(h)	Bent feed strips.
	(i)	Bent trigger arm.
	(j)	Bent or kinked actuator spring.
	(k)	Cartridge stop lost.
	(l)	Stripping finger bent up.
	(m)	Bent feed piece spring.
V. *Bad Ammunition:*	(a)	Reloaded ammunition.
	(b)	Old ammunition.
	(c)	Misfire.

When a stoppage occurs, the first thing to do is to cock the gun, if possible, clear the ejection opening and feedway, load if necessary, and continue firing. If the trouble recurs, one of the causes mentioned above must be looked for and corrected. In rare cases some accident, such as a broken firing pin may cause a serious jamming of the mechanism such that the piece cannot be cocked. This can usually be overcome by placing a screw driver or other blunt instrument in the notch on the left hand side of the breech-block, and tapping the breech-block forward. This will free the jammed parts, and if the breech-block is then held forward by the same instrument while the cocking handle is drawn to the rear, the jam can be freed.

The following table classifies stoppages and jams according to symptoms and gives the immediate action necessary to continue the gun in action:

Symptom	Immediate Action	Cause	Purposes
Gun stops with breech-block half open, empty shell in front of breech-block, or, gun stops with block half open, no cartridge in front of block. Strip only partly fed over.	Cock the gun, clear ejection opening, and screw down regulator 5 points. If this fails, repeat until gun functions properly.	I (a)	Screw out regulator 5 points beyond proper setting.
	If action given above fails, look at ejector cap, and, if necessary, assemble it properly.	I (h)	It is not advisable to set this up.
	If ejector is all right, change barrels; if this fails,	IV (f) III (g)	Not to be set up.
	Change actuators.	III (f)	Use an actuator with cup on end reamed out.
Gun fires automatic when set at "R."	Continue firing, and after the engagement apply the corrections given above, or in case IV (i) bend up trigger arm.	I (a) I (h) IV (f)	Use a trigger with arm bent down slightly.
		III (f) also IV (i)	
Gun stops with two cartridges, one live and one empty, wedged in receiver.	Put in new extractor and spring.	II (d) IV (d) IV (e)	Put in a broken extractor spring
	Or if gun has been racing, screw out gas regulator 5 points.	I (b)	Do not set this up.

Symptom	Immediate Action	Probable Cause	Method of Producing the Malfunctions for Instruction Purposes
	If this fails, or if gun has not been racing, examine ejector.	II (f)	Leave out ejector. (It is not advisable to set this up by leaving out ejector spring.)
	If action given above fails, change breech-block.	IV (c)	Use a worn breech-block.
After gun is loaded, it fires one shot and stops half open with an empty cartridge in front of block. Gun cannot be cocked without lifting feed piece.	Pull back cocking handle as far as possible, at the same time raising feed piece. Cock gun, holding up feed piece. Withdraw strip and insert it properly.	I (c)	Load with end of feed strip under the front feed guide.
The gun stops with breech-block half closed, and a live cartridge in front of block. Cartridge is badly stripped or torn in front, and bullet may be partly driven in.	Cock the gun, clear ejection opening, see that gun is fully loaded, continue the fire. After the engagement, see to the resizing of feed strips.	I (d)	Use a loose feed strip.
	Replace cartridge stop.	IV (k)	Do not set this up.
Gun stops firing with breech-block closed.	Count five slowly to allow time for a hang fire, then cock gun and continue firing.	V (c)	Use a dummy cartridge in feed strip.
	If trouble recurs, if primer is only slightly dented by firing pin, change barrel.	I (e) I (f)	
	If this fails, change firing pin.	IV (g)	Use a short firing pin.

	Let down mechanism slowly to see if fermeture nut locks completely. If not, change fermeture nut.	I (g)	Put some brass filings or wet dirt in fermeture nut.
	If primer is not struck by firing pin, examine handguard to see if it is properly assembled (front stiffening piece above lug on barrel. If handguard is bent or dented, remove it and continue firing.	I (i) Note—In this case the m. f. usually occurs on 1st shot, as well as on all succeeding ones.	
	Examine firing pin, if necessary.	II (c)	
	Look at corner of slot in breech-block. If broken, change breech-block.	II (e)	
Gun ruptures cartridge case about $3/4$-inch from the base.	Clear the chamber with defective cartridge extractor. Unscrew locking screw 3 turns, then with dismounting wrench, tighten the locking nut as much as possible. Bind locking nut in new position by tightening locking screw. Continue firing. At first opportunity change locking nut.	III (a)	Use a feed strip containing one or more cartridges which have had a groove filed around them $3/4$ of an inch above the base.

	If the above action does not remedy the defect, change barrel.	IV (b)	
	If this is not effective, change breech-block.	III (d)	
	Change fermeture nut.	III (c)	
	If all the above remedies fail, the receiver is strained and the gun should be turned in for repairs at the earliest opportunity.	III (b)	
Gun stops with breech-block partly shut, underrunning cartridge.	Cock gun, remove cartridge, push feed strip over until cartridge stop projects, continue firing. When opportunity occurs, resize feed strips.	I (d)	Use a loose strip. (This will not always cause this malfunction.)
	Correct the fault by bending stripping finger down carefully.	IV (l)	Do not set this up.

SCHEDULE OF INSTRUCTION:

The following program of instruction was developed by the authors during their experience as machine gun instructor of troops on the Mexican Border in 1916-17. The program is intended to give the organization as thorough a knowledge of the mechanism as can be imparted in the limited period of two weeks. It should be supplemented by target practice and by instruction in minor tactics.

PROGRAM OF INSTRUCTION FOR COMPANIES ARMED WITH THE AUTOMATIC MACHINE RIFLE, CALIBER .30, MODEL OF 1909

(Benet-Mercié)

1. The course will consist of one week in the study of the mechanism of the piece, and one week spent in exercises which involve firing. The hours of instruction are from 8 A. M. to 11 A. M., and from 1:30 P. M. to 4 P. M.

2. The program for the first week is as follows:

MONDAY FORENOON:

Lecture on the history and development of machine guns and their mechanism.

Mechanism: Principles of the gun explained briefly by the instructor.

Work for Class: Disassembling and assembling, and nomenclature.

MONDAY AFTERNOON:

Instruction in nomenclature by the company officers. When each member of the class can pick out every part mentioned in the alphabetical list of components of the rifle, and can name any part of the gun picked up at random by the instructor, he will be considered sufficiently instructed in nomenclature.

TUESDAY FORENOON:

Lecture on the characteristics which affect the tactical employment of Machine Guns.

Mechanism: Instruction in the proper method of assembling and disassembling the rifle in which the precautions necessary to prevent the abuse of the rifle in taking it down and putting it together are carefully explained.

Work by Class: (Blindfolded) Nomenclature of parts, including contents of gunner's pouch and spare parts case. Assembling and disassembling. Each member of each squad will also be required to practice cocking the piece while in the firing position on the ground.

TUESDAY AFTERNOON:

Assembling and dissassembling blindfolded.

WEDNESDAY FORENOON:

Lecture: The mechanism of the rifle and the functions of each part.

Mechanism: Instruction of the class by officers and non-commissioned officers of the company in the mechanism of the rifle, two guns to be used by each class. One gun will be assembled and placed on the table on field mount, and the other one will be disassembled and the parts laid on the table. As the movement of the gun in firing is explained, it will be illustrated on the assembled gun with dummy cartridges, and the student will be required to move each part of the disassembled gun in such a manner as to simulate the movement of the corresponding part in the operation of the gun.

WEDNESDAY AFTERNOON:

Functioning of the piece, as above, continued.

THURSDAY FORENOON:

Lecture: Jams, malfunctions and stoppages. Their prevention. Care and preservation of the piece.

Work for Class: Replacing of parts. Effort should be made to learn thoroughly the most expeditious method of replacing the following parts.

Barrel
Locking nut
Fermeture nut
Breech-block
Firing pin
Extractor
Extractor spring
Feed piece
Feed piece spring

THURSDAY AFTERNOON:

Replacing of parts the same as for Thursday forenoon. At the completion of this practice the time of each member of each squad in performing each of these operations will be averaged by the company commander to determine the relative proficiency of each squad in this work. For tests of this character the gun squad should be in the firing position on the ground.

FRIDAY FORENOON:

Lecture: Examination of the gun to verify its readiness for action.

Mechanism: After the lecture each gun will be gone over in the manner indicated and all defects will be corrected.

Work for Class: Drill in loading the piece. For this drill, dummy cartridges will be used and an effort will be made to develop the greatest possible speed and accuracy in placing the strips in the gun. All members of the squad will be practiced in the position of loader.

FRIDAY AFTERNOON:

The resizing of feed strips and instructions in the care and preservation of the guns.

SATURDAY FORENOON:

Lecture on the indication and recognition of targets. Practice in applying principles embodied in lecture, employing groups of targets concealed in the terrain near the school.

SECOND WEEK

MONDAY FORENOON:

Adjustment of each piece by actual firing and recording all data as to gas setting, etc., for each barrel.

MONDAY AFTERNOON:

Adjusting of telescopic sights.

TUESDAY FORENOON:

This period will be spent in familiarizing the entire company with the actual firing of the gun. This will be done by allowing each man to fire half a strip (15 shots).

TUESDAY AFTERNOON:

Loading practice. Each member of the company is required to act as loader and load five (5) strips each loaded with three cartridges in each end.

WEDNESDAY FORENOON:

Loading practice continued.

WEDNESDAY AFTERNOON:

Exercise in malfunctions, stoppages, and jams. Each member of each Gun Squad will be required to act as gunner and fire a gun which has been previously put out of order in such a manner as to produce a predetermined malfunction, stoppage or jam, the nature of which will be unknown to the man firing the gun. He will be required to get the gun in action and fire a given number of shots, some of which defective ammunition will be used, in the shortest possible time.

THURSDAY FORENOON AND AFTERNOON:

Malfunctions, stoppages and jams same as for Wednesday Forenoon.

FRIDAY FORENOON:

Exercises to determine the size of shot group made by each gunner. For this purpose targets 1, 2, 3, and 4 will be used. Exercises in traversing and searching.

FRIDAY AFTERNOON:

Practice in bringing up guns and going into action in the shortest possible time. Practice with the night firing box. Practice to determine how many aimed shots each squad can fire in one minute.

SATURDAY FORENOON:

This period will be devoted to a firing problem at designated silhouette targets concealed in positions adjacent to the instruction range. Each squad will be given one full box (300 rounds) of ammunition. The company will be allowed to reconnoiter the terrain for the purpose of locating the hostile line, after which the company will be required to come up to the firing point from cover, go into action, and fire all ammunition allotted for the problem. After this exercise the hits on the targets will be counted and the instruction of the company will end.

CHAPTER II

The Lewis Automatic Machine Gun.

This weapon is the invention of an American, Colonel I. N. Lewis, Coast Artillery Corps (now retired). It attained great prominence immediately after the beginning of the European War, when several thousand weapons of this type were purchased by some of the belligerent nations.

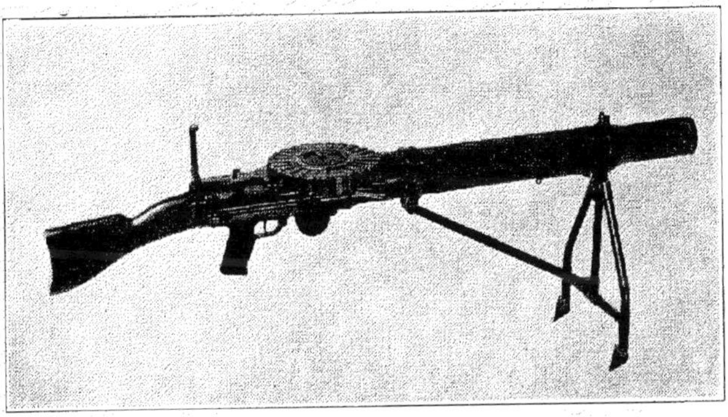

The gun is of the light air-cooled, gas operated type, fed from a magazine. A very desirable feature of this gun is its distinctive air-cooling system, which cools the gun almost as effectively as a water jacket, without the attending disadvantages of increased weight, difficulty of water supply, and production of steam which are found in most water-cooled guns. This gun also has the advantage of comparative simplicity and ease of assembling and disassembling. It weighs but 26 lbs. and can easily be handled by one man. In case of necessity this gun can be transported from one position to another like an ordinary shoulder rifle. The ease with which the different groups of the mechanism can be independently disassembled renders the reduction of jams very rapid and sure.

Several hundred Lewis Machine Guns were issued to U. S. Troops on the Mexican border during the summer of 1916. These guns are all chambered for the Mark VII .303 British ammunition, and is the type of Lewis Gun covered by the description in this chapter. The appearance of the gun is shown in the Plate.

DESCRIPTION OF THE MECHANISM.

In the accompanying sectional view of the gun, all of the parts are numbered and named for convenient reference in connection with the following description of the operation of the gun.

Principle of Operation.—The gun is operated automatically by trapping a small portion of the live powder gases before the bullet leaves the muzzle, and causing this portion of gas to impinge against the head of a free moving piston, which is thus driven back against the force of a spring and is returned by this spring when the force of the gases is spent. The motion of the piston is utilized to unlock the breech bolt, eject the empty shell, feed in a new cartridge, relock the breech bolt, and fire.

To Put Gun into Action.—When a loaded magazine is dropped into place over the magazine post, and the charging handle pulled to the rear as far as it will go and then released, the gun is ready to be fired by a pull on the trigger.

Single Shots.—When the trigger is pulled once and very quickly released a single shot is fired and this may be repeated at will until the magazine is empty.

Automatic Fire.—Within the limits of the magazine capacity the gun continues to fire so long as the trigger is held back and stops firing whenever the trigger is released. It follows therefore that the operator may at will fire shots either singly or in groups of two, three, four, or of any number up to the full magazine capacity of forty-seven cartridges. Continued pressure on the trigger therefore results in automatic fire, which need be interrupted only by the four-second intervals required to replace emptied magazines by loaded ones.

Cooling System.—No moving parts are employed. The method of cooling is as follows: Closely fitting the steel barrel is a cylindrical radiator of aluminum having deeply cut longitudinal grooves throughout its length and circumference. Over this aluminum radiator there is a thin tubular steel casing, the muzzle end of which extends at reduced diameter beyond the end of the radiator and barrel. The radiator and tubular casing together with the barrel mouthpiece constitute the entire cooling system. The specially shaped barrel mouthpiece screwed to the end of the barrel serves the double purpose of firmly securing the radiator in place and of so directing the powder blast at each discharge as to induce a current of cool air in at the rear end of the radiator casing, through the longitudinal grooves of the radiator, and out at the front end of the radiator casing. The tubular steel casing serves to confine the blast of cool air and thus keep it in contact with the aluminum of the radiator. The high specific heat, the great heat conductivity, and the low specific gravity of aluminum, combined with an exceeding simple and durable construction, produce a cooling system for the gun that is practical and efficient, without rendering it too heavy or bulky for general field service.

Action of the Mechanism.—Starting with the gun in the "ready-to-feed" position, as shown in the Plate, when the trigger (30) is pulled and held back the action is as follows:

Actuated by the force of the mainspring (55) acting through the gear (52) and the rack (72), the operating rod moves forward, carrying with it the bolt (37).

As it moves forward the front top edge of the bolt (37), striking the lower edge of the rim of the cartridge resting in position in the feedway, drives the cartridge forward and downward into the chamber.

At the end of the forward movement, the resistance lugs on the rear end of the bolt (37) emerge from their guide slots into the space in the receiver (26) known as the locking recess. The bolt (37) is now free to turn and lock, which movement is accomplished by the action of the operating post (m) against the shoulder of the cam slot in the bolt. As the bolt closes behind the cartridge in the chamber the extractors (45) take their grip upon the rim of the cartridge case.

During the forward movement of the bolt and operating rod, the magazine has been held by the rebound pawl (29), and has not moved.

The feed operating arm (34), actuated by the feed operating stud (31) carried by the bolt (37), has been returned to the right in position to take the next cartridge.

When the bolt has fully turned into the locked position the striker (50) is free to drive forward and fire the cartridge.

When the cartridge is fired, no action takes place in the gun mechanism until the bullet reaches the gas port in the barrel (76).

While the bullet is passing from the gas port to the muzzle, a small portion of the live powder gas enters from the bore through the gas port into the gas regulator cup (84), where it deposits any solid matter that it carries with it. The clean gas expands through an aperture in the gas regulator cup (84) and a corresponding aperture in the wall of the gas chamber (82), against the head of the piston (79).

The force of the gas drives the piston to the rear against the force of the mainspring and during this movement the following actions take place.

The piston (79), acting through the rack (72), on the gear (52), winds up the mainspring (55).

The operating post (m), acting on the side of the cam slot in the bolt (37), rotates the bolt sufficiently to unlock it, then carries the bolt straight back to the rear.

The extractors (45) extract the empty cartridge case, which is thrown out by the ejector (21), actuated by the feed operating stud (31) striking against its rear end.

The feed operating stud (31), carried by the bolt (37) and acting on the sides of the channel in the under side of the feed operating arm (34), moves the arm to the left, carrying a cartridge from the magazine under the cartridge guide spring (40) and into the feeding position in the feedway on top of the receiver (26).

The feed pawl (35), carried by the feed operating arm (34) and acting on the outside wall of the magazine pan (43), carries the magazine through a partial revolution sufficient to bring the next cartridge in position. The magazine is held in its new position by the stop pawl (28) and the rebound pawl (29) and does not move during the forward movement of the other parts.

The foregoing operations are all completed by the impulse given to the piston by the direct action of the gas. The last of this impulse is expended when the operating rod and bolt come to rest against the butt tang (5) at the extreme end of their movement to the rear. The forward movement then begins and the cycle of operations is repeated for each shot until the magazine is empty, when the parts stop at the end of the forward movement, with the bolt locked behind the empty chamber.

During the firing, if at any time the trigger is released before the magazine has been emptied, the gun stops at the beginning of the forward stroke, in the "ready-to-feed" position. In this position the action is open and the chamber empty, but firing is resumed on pulling the trigger.

A Short Discussion of the Common Difficulties Likely to be Experienced With Each Part is Given Below

Barrel Mouthpiece.—This part should not be disassembled except when it is necessary, and when it is reassembled it should be screwed on as tightly as possible, as otherwise it may become loosened by the vibration of firing and be blown off and lost. When this happens it can be detected by the characteristic change in the noise of the firing.

Gas Chamber Band.—This piece is tapered to fit the taper of the barrel, and care should be taken in assembling it to put it on with the taper in the right direction. The best way to do this is to try the band on before the barrel is put into the radiator to see if the hole in the band can be made to register with the gas port in the barrel. Holding the band in the same relative position, it should then be placed in its recess in the radiator, and the barrel should then be inserted.

Gas Chamber.—The threads of the gas chamber may be stripped by careless assembling.

Gas Cylinder.—The gas cylinder may have the threads stripped by careless assembling. In this case it will blow loose from the gas chamber, and stop the action of the gun. At the same time it may work back into the receiver, and when the defect is discovered trouble will be encountered in unscrewing the receiver from the barrel group.

All of the above mentioned troubles can be avoided if the barrel group is never disassembled except when it is absolutely necessary.

Operating Rod.—The striker may become broken or damaged. In this case, use the spare operating rod until opportunity can be found to remove the striker fixing pin and substitute a new striker for the damaged one. The operating rod may become bent. Usually when this happens it can be straightened. One or more teeth may become broken off the rack, or the sear notch may have the edges broken or chipped.

Barrel.—The barrel has at its rear end an annular lip for supporting the rim of the cartridge. The fact that the lip is cut away at the top to afford a feedway, and at the side to give clearance for the extractors, gives the rear end of the barrel a rather ragged appearance. Persons unfamiliar with the gun have been known to mistake this for a defect in manufacture, and in some cases attempts to correct the supposed defect by filing off these lips have been made, thereby ruining the barrel. These lips have been known to break off during the firing. In this case the scraping of the cartridges on the edge of the barrel interferes with the feeding.

Radiator.—This is made of aluminum which has a larger coefficient of expansion than the steel of the barrel. Consequently an attempt to disassemble the barrel group in very cold weather will sometimes fail owing to the fact that the contraction of the aluminum radiator has caused it to grip the barrel. This can be overcome by heating the radiator slightly.

Clamping Ring. In assembling this part, care should be taken to see that it is put on with the high part of the front sight to the rear.

Receiver.—In many cases it has been found that the conformation of the guiding lips on the receiver is such that the cartridges will fail to feed, tipping up and striking in the throat of the feedway. This can also be caused by a weak cartridge guide spring.

This defect can be corrected by changing the conformation of the guiding lips of the receiver, but the change should never be undertaken except at an Arsenal by workmen thoroughly familiar with the method.

Safety.—The safeties should not be removed except in case of necessity, as they can easily be broken in the operation of removing or replacing them.

Mainspring.—Occasionally the mainspring breaks. When this occurs it should be replaced by a new one.

Gear Casing.—By abuse, the sides of the gear casing may become bent in so as to bind and interfere with the action of the spring. This defect can be overcome by carefully spreading the sides of the gear case.

Trigger Mechanism.—Occasionally the trigger mechanism becomes clogged by dirt or by powder grains from a broken shell. When this happens, the gun will not stop firing when the finger is removed from the trigger. In this case the firing can be stopped by pushing the trigger forward. At the first opportunity after such an occurrence the trigger mechanism should be dismounted and thoroughly cleaned and oiled.

Feed Operating Arm.—Occasionally a burr will occur on the stud at the rear end of the operating arm. This will stop the fire of the gun. The burr should be carefully removed with a fine file. After this accident has once happened it will recur with persistent regularity. It is therefore best to obtain a new arm at the first opportunity.

Lewis machine gun on motorcycle mount

Bolt.—Occasionally a bolt will crack from the shock of firing. Sometimes the edges of the cam slot in the bolt become chipped or broken. The obvious remedy is to use the spare bolt.

Ejector.—The ejector frequently becomes badly burred at the back end. It should be watched, and when this occurs, the burrs should be carefully removed with a fine file, taking care not to alter the original shape.

Ejector Cover.—This should not be removed at drills, or at any time except when necessary, as the frequent removal of this part causes its seating in the receiver to become worn so that the ejector cover jumps out during firing, stopping the action of the piece.

When the seat for ejector cover has become worn in this manner, the rear end of the seat should be undercut, so as to overcome the tendency of the ejector cover to jump out. This can be done with a small scraper and should be attempted only by a skilled mechanic. If there are no suitable tools on hand for this work a temporary remedy may be obtained by burring the rear edge of ejector cover seat slightly.

Feed Cover.—Care should be taken that the cartridge guide arm on the feed cover does not get bent as this will interfere with the feeding. If the arm is bent up, it will strike the top row of cartridges in the magazine, preventing the magazine from rotating. Difficulty will be experienced in cocking the gun, and on removing the magazine a mark can be seen on the first cartridge in the top row, showing where it has been struck by the cartridge guide arm.

The remedy is to bend the cartridge guide arm carefully into its correct position, comparing it with a feed cover which is in good condition. The bending of the cartridge guide arm should be done while the feed cover is in place on the gun, in order to avoid distorting the front extension of the feed cover, thus rendering it difficult to assemble to the receiver.

Magazines.—Magazines may become distorted from being dropped, or from using undue force in placing them on the gun or removing them from it. A distorted magazine will cause failure to feed. The only remedy is to replace it.

Buttstock.—In placing the gun in the firing position, care should be taken not to strike the buttstock on the ground, because a comparatively light blow on the toe of the stock will break the butt tang screw.

EXAMINATION OF THE GUN TO VERIFY ITS READINESS FOR ACTION:

If the gun is periodically examined and all noticeable faults are corrected, malfunctions on the firing line will be reduced to a minimum. The examination is best conducted as follows:

Dismount the gun, clean and oil all parts.

Guard.—Examine trigger mechanism by pulling and releasing trigger. If sear does not work freely dismount and clean all parts thoroughly and repeat the test, taking care to assemble front end of sear in its proper notch in trigger. If it still fails to return, use the spare sear. This trouble may be due to the fact that the sear is a little too long. In this case the defect may be corrected by filing.

Stock.—Tighten butt tang screw if possible.

Feed Cover.—If rear sight bed spring screw is loose, tighten it. Dismount cartridge guide, examine cartridge guide spring. If weak or deformed, replace it. Dismount magazine pawls. Examine pawl spring. If cracked or deformed, replace it. Replace pawls and see if they work freely. (Note—Pawls are very hard, and rarely ever wear enough to affect the operation.) Examine cartridge guide arm on feed cover. If it has been bent or twisted, or if the horns are spread or bent, correct the defect, using a copper hammer and copper drift. Defects of this nature can best be detected by comparing the feed cover with one known to be correct. No change should be made unless the defect is plainly noticeable and the result to be accomplished is understood. The change should be made while the feed cover is in place on the receiver as otherwise the bending may take place at the wrong point, and difficulty will be encountered in assembling the feed cover to the receiver.

Feed Operating Arm.—Examine arm in place to see if it is bent up in the middle. If so, correct the defect. Examine the lug under the rear end of arm to see if a burr is being raised at this point. If so, remove the burr carefully with a dead smooth file. If a burr once starts here it will recur, and it is suggested that if this trouble is detected steps should be taken to obtain a new arm.

Examine feed pawl spring by comparing it with a new one. If it is deformed or weak, repair it.

Bolt.—Examine face of bolt to see if it is cracked. If so, replace it. Examine inside of cam slot for roughness. If any is detected, smooth it off with a file or emery cloth. In assembling, be sure that this part is thoroughly oiled, as the proper working of the gun is largely dependent on keeping this part from becoming burred or rough.

Operating rod.—Insert striker in bolt and test to see if it projects through far enough to fire a primer. If not, replace striker. Examine operating post. If it is badly worn smooth it up or use spare operating rod. Examine sear notch in operating rod. If chipped or broken, use spare operating rod.

Gear Casing.—Dismount and examine the parts to see

(a) That mainspring is not broken.

(b) That spring is properly assembled to collet.

(c) That sides of gear case are not bent in so as to impede the action of the gear. Test gear stop to see that it acts freely and that spring is all right.

Receiver.—Examine for burrs. Look particularly at inside of receiver above feedway. If any burrs are found, remove them with a file.

Barrel Group.—Ordinarily the barrel group should not be dismounted, but when it becomes necessary, the following points should be observed. See that barrel where it enters radiator is thoroughly cleaned and is oiled lightly. See that barrel mouthpiece is screwed on as tightly as possible, remembering that this is a left handed thread. Examine threads on gas chamber, gas chamber band, and gas cylinder to see that they are not stripped. See that gas cylinder is not burred at the rear end by carelessly allowing it to strike against the radiator casing in assembling.

Magazines.—Test the magazines by spinning them (empty) on the loading handle. Examine the rim to see that it is not deformed.

Final test.—After assembling the gun, draw back to charging handle, then pull the trigger, to see that the mechanism works correctly. If difficulty is experienced in cocking the piece, it may be due to one of the following causes:

(a) Ejector cover not seated. This will cause the feed operating arm to bind on the projecting end of ejector cover.

(b) Operating stud not in cam groove in feed operating arm.

(c) Magazine pawls assembled in wrong posts. (Stop pawl in place of rebound pawl, and vice versa.)

(d) Failure to oil cam surface inside of bolt when this surface has become worn or roughened.

METHOD OF HOLDING THE GUN:

The proper method of holding the gun is shown in the photograph. The gunner lies squarely behind the gun, and does not incline his body to the left as with a shoulder rifle.

He grasps the guard firmly with his right hand, with the index finger on the trigger, and the thumb around the guard so as to be out of the path of the charging handle.

Proper method of holding the gun to insure steadiness

The right elbow rests on the ground. With left hand he supports the toe of the stock, resting the elbow on the ground for steadiness.

To LOAD:

Grasp the magazine firmly with both hands as shown in the photograph, holding it with the arrow to the front, and place it on the magazine post, turning it slightly in each direction to be sure it is seated. At night when the arrow cannot be seen, the right thumb may be placed lightly on the magazine latch to determine the proper position. In placing the magazine on the post the magazine latch must not be pressed in, as this will release the magazine center and may cause one of the cartridges to drop and cause a malfunction.

To UNLOAD:

When firing is stopped and a partly empty magazine is removed, a cartridge remains in the feedway. When circumstances permit of such a procedure it may be removed by firing. If it is not desirable to unload by firing, proceed as follows:

(a) Remove the magazine.

(b) Holding cocking-handle with the right hand, press the trigger with the left hand, ease the cocking handle forward with just sufficient force to push cartridge from slot into the body of the receiver.

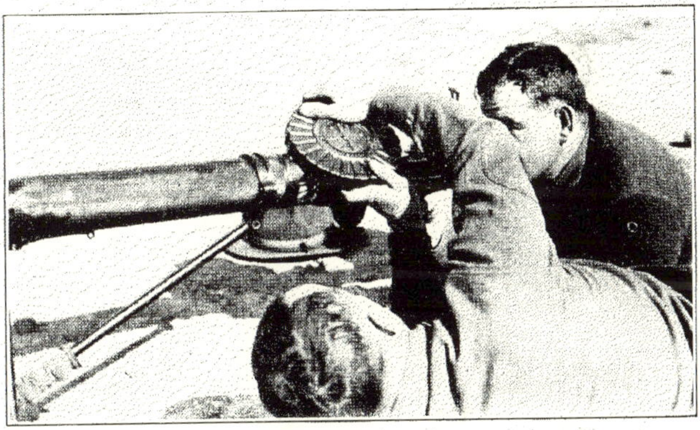

Proper method of loading

(c) Pull back the cocking-handle until the notch on rack engages with the nose of the sear.

(d) Raise the safety.

(e) With the point of the bullet of a spare round pushed through the slot on top of the body, press against the rim of the cartridge on the left side, and draw cartridge out through ejection opening.

(f) Lower safety catch, press the trigger.

POINTS TO BE OBSERVED BEFORE, DURING, AND AFTER FIRING

Points to be observed before going on the firing line for target practice or action.

(a) Examine the gun to see that no part is broken or damaged and that the mechanism works freely.

(b) Oil all working parts, especially the cam slot in the bolt and the operating post. The threaded part of the gas regulator cup may be oiled lightly.

(c) See that surplus oil is removed and the bore is clear.

(d) See that gas regulator cup is properly adjusted. Ordinarily the small gas port should be used but in cold weather or at night the large port should be used. The large gas port should also be used when sand is blowing.

(e) With the gun assembled, test the weight of the mainspring by applying spring balance to the charging handle. The weight required to hold the charging handle steady just after it has been started to the rear, should be from 8 to 10 pounds.

(f) Place an empty magazine on the gun, and work the charging handle to see that feed mechanism is working correctly.

(g) Before filling, examine each magazine to see that separator pins are not broken or bent, and spin it on loading handle to be sure that it is not distorted and works freely.

(h) See that magazines are carefully filled. If cartridges jam while passing into magazine, it should be emptied and examined to ascertain the cause.

(i) See that all spare parts and tools are in their proper place and available for use, and the oil can is full of oil.

(j) See that gun, magazine, and spare part cases are properly secured to avoid loss or damage in transit.

Points to be observed during firing:

(a) Raise the safety on the side opposite to the charging handle. This will prevent gas and oil from flying into the gunner's face.

(b) Take advantage of every available opportunity to reoil the working parts.

(c) During a cessation of fire, replace a partly empty magazine on the gun with a full one.

(d) Refill empty magazines at every opportunity.

Points to be observed after firing:

(a) See that gun is unloaded.

(b) See that bore and chamber, also gas cylinder and operating rod are well oiled as soon as possible after firing has ceased.

(c) See that mainspring is eased by returning charging handle to forward position.

(d) See that the gun is thoroughly cleaned without delay on returning to camp or barracks. All parts of the mechanism, as well as the magazines, must be examined at the same time, to see that they are in good order.

(e) See that the barrel of the gun is cleaned once a day for at least ten (10) days after firing. This is necessary to prevent corrosion from the effect of the smokeless powder used.

JAMS AND MALFUNCTIONS:

The most frequent cause of malfunctions with this gun is wear or deformation of the rim of the magazine, which causes a failure to feed. To minimize malfunctions from this cause, it is necessary to exercise the utmost care to avoid damage to the magazines. In any mechanism with as many requirements and limitations as a machine gun, there are necessarily a number of things which may cause trouble. The more common of these causes are listed in the following table:

I. *Improper Manipulation:*

(a) Wrong spring tension.
(b) Magazine improperly placed on gun.
(c) Magazine improperly loaded.
(d) Gas regulator cup partly unscrewed.
(e) Magazine pawls wrongly assembled.
(f) Operating stud left out in assembling.
(g) Operating stud not in slot of feed operating arm.
(h) Ejector cover not properly assembled.
(i) Charging handle not properly assembled.
(j) Barrel mouthpiece not screwed on tight.
(k) Dirt in trigger mechanism.
(l) Dirt in chamber.
(m) Failure to oil operating post and cam slot in bolt.

II. *Breakages:*

(a) Main spring.
(b) Bolt.
(c) Striker.
(d) Butt tang screw.
(e) Top extractor.
(f) Lips on rear end of barrel.
(g) Teeth on rack.
(h) Sear notch.

[II. *Worn or deformed parts:*

(a) Sides of Gear Casing bent in, binding the gear.
(b) Burr on rear lug of feed operating arm.
(c) Burred ejector.
(d) Receiver lips incorrect, causing tip-up.
(e) Worn or deformed rim on magazine.
(f) Cartridge guide arm bent.
(g) Cartridge guide spring weak.
(h) Feed pawl spring weak.
(i) Feed operating arm bent up in the middle.
(j) Excessive headspace, causing ruptured cartridge.
(k) Top extractor weak, allowing it to become pushed up and back slightly out of its seating.
(l) Ejector cover seating in receiver worn at rear end.

IV. *Defective ammunition:*
- (a) Misfire.
- (b) Large head on cartridge case.
- (c) Battered or deformed ammunition.

The following table classifies stoppages and jams according to symptoms and gives the immediate action necessary to continue the gun in action in each case.

The positions in which the gun may stop are shown in the accompanying photographs, and are as follows:

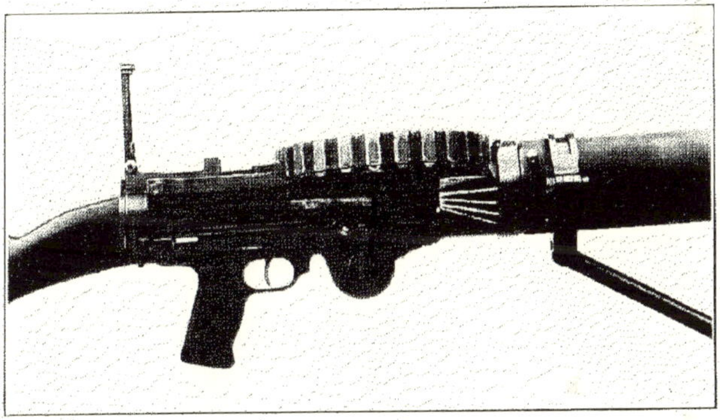

First characteristic position of the charging handle

1st position:	Charging handle forward.
2nd position:	Charging handle back less than the length of a cartridge.
3rd position:	Charging handle back more than the length of a cartridge but not as far back as position No. 4.
4th position:	Charging handle 1 inch forward of cocked position.
5th position:	Charging handle stops to the rear of cocked position.

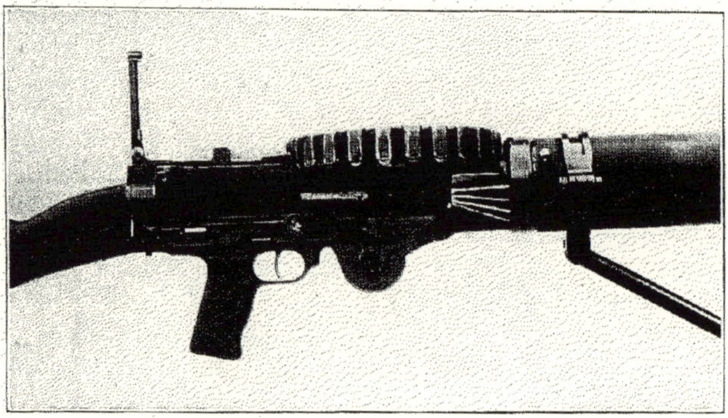

Second characteristic position of charging handle

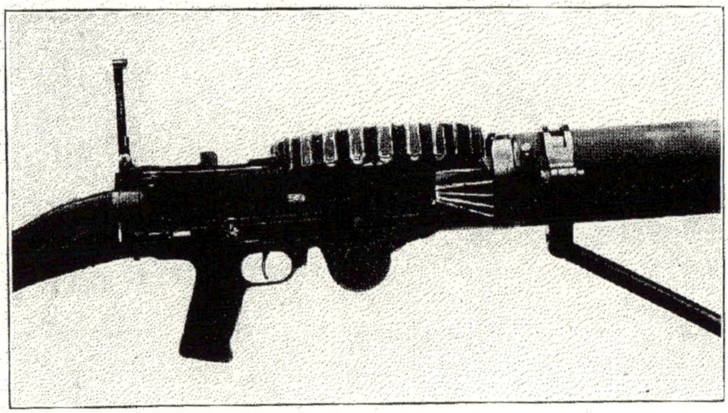

Third characteristic position of the charging handle

Fourth characteristic position of the charging handle

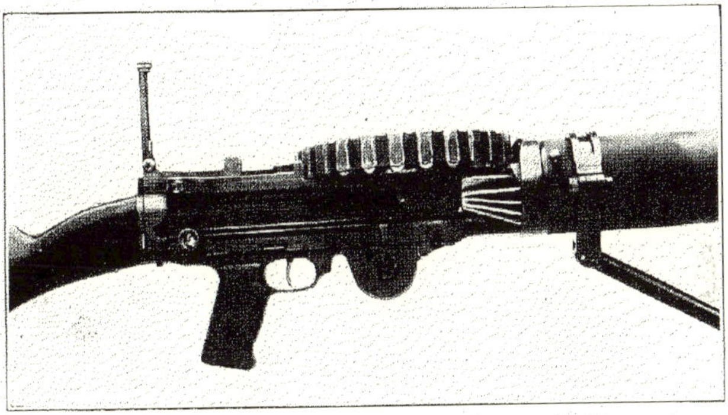

Fifth characteristic position of the charging handle

Symptom	Immediate Action	Probable Cause	Method of Setting Up for Purposes of Instruction
Gun stops in 1st position.	Try magazine to see if it will rotate to the left.		
	(1) If so, it is empty. Change it.	Empty Magazine	Put an empty magazine on gun.
	(2) If not, pull back charging handle and continue firing.	I (c)	Leave an empty space in magazine.
		IV (a)	Place a dummy cartridge in magazine.
		IV (c)	Place a damaged round in magazine.
	If trouble recurs:		
	(1) If empty case comes out of gun when handle is pulled back, look at gas regulator cup, and if partly unscrewed, correct it.	I (d)	Unscrew gas regulator cup one-half turn.
	(2) If live round comes out, note tension of main spring. [illegible] spring is weak, increase the tension.	I (a)	Not to be set up.
	If it is broken, use the spare one.	II (a)	Not to be set up.
	If spring is correct, examine striker. If it is broken or bent, use spare operating rod.	II (c)	Use a broken striker.
	If striker and spring are correct, wipe out chamber.	I (e)	

	(3) If nothing comes out of chamber when handle is pulled back, change magazines.	III (e)	Use a deformed magazine.
	If this fails, take off feed cover and see whether or not operating stud has been left out. If so, replace it.	I (f)	Leave out operating stud in assembling.
	If operating stud is correct, replace feed pawl and feed pawl spring.	III (h)	Use a weak feed pawl spring.
	(4) If difficulty is experienced in cocking the piece, oil cam slot in bolt, and continue firing.	I (m)	Not to be set up.
Gun stops in 2nd position.	Pull back charging handle and examine the ejected cartridge.		
	(1) If it is deformed or battered, continue firing.	IV (c)	Use a battered cartridge.
	(2) If it has the front part of a ruptured case telescoped onto it, continue firing.	III (j)	File a groove around cartridge case 1½ inches from base.
	(3) If it is perfect, examine top extractor. If it is up out of its proper seating, use spare bolt.	III(k)	Push top extractor up and back slightly out of its proper seating after assembling gun.
	If top extractor is correct, there is probably the front part of a ruptured case in the chamber. Remove it with the defective cartridge extractor, then continue firing.	III (j)	File a groove around cartridge case one-half inch from base.

Symptom	Immediate Action	Probable Cause	Method of Setting Up for Purposes of Instruction
	If rupture of cartridge cases recur, change bolt.		
	If this is not effective, change barrels at the first opportunity.		
Gun stops with handle in 3rd position.	Pull back charging handle.		
	(1) If a considerable amount of friction is encountered, remove magazine and feed cover, and with the point of a bullet push ejector cover into its proper place. Assemble parts and continue firing.	III (e)	Not to be set up.
	(2) If no unusual difficulty is encountered in cocking the piece, take off magazine and remove cartridge from feedway.	IV (b)	Use a cartridge with a large head.
	Replace magazine and continue firing.		
	(3) If trouble recurs, examine top extractor, replacing it if it is broken.	II (e)	Not to be set up.
Gun stops in 4th position.	Pull back charging handle and continue firing. If this fails, pull back charging handle, take off	III (d) III (g) III (i)	Use a weak cartridge guide spring.

	magazine, push down point of cartridge which caught in throat of feedway, replace magazine, continue firing. At the first opportunity change cartridge guide spring.		
Gun stops in 5th position.	Try charging handle to see if tension of mainspring is gone. If so, use the spare mainspring.	II (a)	Not to be set up.
	If charging handle is fixed and cannot be moved by hand, remove magazine, push cartridge that is in feedway down into receiver and remove it. Take off feed cover and lift up feed operating arm which will free operating stud and allow charging handle to go forward. Increase tension on mainspring to about 15 pounds. If a file is available, smooth off the burr on rear lug of operating rod. Continue firing. At the first opportunity, obtain a new feed operating arm.	III (b)	Not to be set up.

SCHEDULE OF INSTRUCTION:

The following program of instruction was developed by the authors during their experience as machine gun instructors of troops on the Mexican Border in 1916-1917. The program is intended to give the organization as thorough a knowledge of the mechanism as can be imparted in the limited period of two weeks. It should be supplemented by target practice and by instruction in minor tactics.

PROGRAM OF INSTRUCTION FOR COMPANIES ARMED WITH THE LEWIS AUTOMATIC MACHINE GUN

1. The course will consist of one week spent in the study of the mechanism of the piece, and one week spent in exercises which involve firing. The hours of instruction are from 8 A. M. to 11 A. M. and from 1:30 P. M. to 4 P. M.

2. The program for the first week is as follows:

MONDAY FORENOON:

Lecture on the history and development of Machine Guns and their mechanism.

Mechanism: Principles of the gun explained briefly by instructor.

Work by Class: Disassembling and assembling, and nomenclature.

MONDAY AFTERNOON:

Instruction by Company Officer in nomenclature.

TUESDAY FORENOON:

Lecture on tactical development of Machine Guns and their importance in War.

Mechanism: Function of each part explained to officers and noncommissioned officers.

Work by Class: (Blindfolded) Nomenclature of parts (Including contents of spare part box).

TUESDAY AFTERNOON:

Assembling and disassembling blindfolded.

WEDNESDAY FORENOON:

Lecture on characteristics of Machine Guns.

Mechanism: Instruction of company by officers and non-commissioned officers on functions of each part, 2 guns to be used by each class. The gun will be assembled and placed on L. F. M., the other will be disassembled and laid on table. As movement of gun in firing is explained, it will be illustrated on assembled gun with dummy cartridge and the student will be required to move each part of disassembled gun in such a manner as to stimulate the movement of corresponding part in the operation of the gun.

WEDNESDAY AFTERNOON:

Functioning of piece as above, continued.

THURSDAY FORENOON:

Lecture: Examination of gun to verify its readiness for action. Points to be observed before, during, and after firing.

Work for Class: Replacing of parts. Efforts should be made to learn the most expeditious method of replacing the following parts:

Extractors
Feed pawl
Feed pawl spring
Cartridge guide lever
Cartridge guide spring
Magazine pawls
Bolt

Operating rod
Receiver lock pin
Gear casing complete
Sear
Charging handle

THURSDAY AFTERNOON:

Drill in replacing parts. Students will explain in detail symptoms and effect of breakage of each part replaced.

FRIDAY FORENOON:

Lecture on Jams, Malfunctions, and Breakages, and on the care and preservation of the piece.

Mechanism: Practice in putting on and removing magazines. For the practice the gun squads will take position on ground as for firing and duties of each man will be performed in turn by each other member of gun squad.

FRIDAY AFTERNOON:

Competitive examination of gun squads in nomenclature, functions of each part, and assembling and disassembling. Results will enter in final figure of merit of each squad.

SATURDAY FORENOON:

Lecture on indication and recognition of targets. Practice in applying principles embodied in lecture, employing groups of targets concealed on the terrain near the school.

SECOND WEEK

MONDAY FORENOON:

Instruction of gun squads in positions and duties of each member of squad. Method of loading, setting sight, and holding piece. In this practice each member of the gun squad will be required to act as gunner, loader, and ammunition man. In position of gunner, each member of squad will fire five single shots at silhouette targets, followed by ten shots from magazine.

MONDAY AFTERNOON:

Each member of organization who is not on a gun squad will be allowed to act as a member of a provisional gun squad undergoing same practice as that prescribed above for gun squads.

TUESDAY FORENOON:

Loading practice. Each member of each gun squad will be required to act as loader while squad fires a string of 12 magazines each loaded with 3 shots. This firing will be done at silhouette targets at 300 yards. Speed and accuracy of shooting will each be given equal importance in this exercise.

TUESDAY AFTERNOON:

Record loading practice. Each squad will be brought up from a designated position and will place gun in position in the hasty intrenchments and fire 12 magazines each loaded with 3 shots at a designated silhouette target in shortest possible time. Each member of each squad will act in each position on the squad. Each squad will thus fire 3 times, each member acting as gunner for one string of 12 magazines. The total time of firing for each squad and total number of hits will be recorded.

WEDNESDAY FORENOON:

Stoppages, Malfunctions, and Jams illustrated. Instructors will set up and illustrate the following jams:

(a) No cartridge in chamber due to empty space in magazine.

(b) No cartridge in chamber, fault in feed.

(c) Misfire.

(d) Hard extraction.

(e) Insufficient gas, partial unscrewing of gas regulator cup or stoppage or orifice.

(f) Misfire, due to weak or broken mainspring.

(g) Misfire, due to damaged or broken striker.

(h) Failure of cartridge to enter chamber—damaged ammunition.

(i) Separation of cartridge case.

(j) Cartridge jammed into empty case which has failed to extract.

(k) Fault in feed due to damaged magazine.

(l) Ejector cover raised.

(m) Weak feed pawl spring.

(n) Broken top extractor.

(o) Piece improperly assembled—feed operating stud left out (Sometimes done when gun is to be abandoned to render it unserviceable).

(p) Failure to feed, due to weak cartridge guide spring.

WEDNESDAY AFTERNOON:

Jams and Malfunctions. Each squad will be required to bring up gun to firing point and fire given number of rounds, with a gun which has been set up to show combination of jams listed above. Firing to be done from hasty intrenchments at designated silhouette targets. Time required to place gun in perfect condition and complete firing will be taken by stop watch and recorded. Each member of squad will be practiced in each position.

THURSDAY FORENOON:

Jams and Malfunctions. Record practice. Each squad will fire 3 strings, each member of squads to act as gunner.

Same jams and malfunctions as those on Wednesday P. M. will be used.

THURSDAY AFTERNOON:

Practice in traversing and searching, using targets No. 1 or No. 2. Practice in indirect fire, using night firing box.

FRIDAY FORENOON:

Practice at target No. 1 to determine size of shot groups. Each member of squad to fire one full magazine. The number shots of each group contained in a 5″ x 8″ ellipse, will be added to the figure of merit for the corresponding squad.

FRIDAY AFTERNOON:

Record practice to determine how many aimed shots each squad can fire in one minute at a designated target. Hits will be counted and number of shots fired and twice the number of hits will be added to figure of merit.

SATURDAY FORENOON:

This period will be devoted to a firing problem at designated silhouette targets concealed in positions adjacent to the instruction range. Each squad will be given six full magazines. The company will be allowed to reconnoiter the terrain for the purpose of locating the hostile line, after which the company will be required to come up to the firing point from cover, and go into action and fire all ammunition allotted for the problem. After this exercise the hits on the targets will be counted and the instruction for the company will end.

CHAPTER III

THE MAXIM AUTOMATIC MACHINE GUN
THE VICKERS MACHINE GUN

As these two guns are identical in principle and, with a few exceptions, are similar in construction, they will be treated together.

The nomenclature followed in this chapter is that given in Ordnance Pamphlet No. 1770 (Handbook for the Maxim Automatic Machine Gun).

The gun is of the heavy type, water cooled, recoil operated, and fed from a belt. The heavy tripod mount gives the gun steadiness and ease of control, and this fact, coupled with the large volume of fire which the belt feed gives, makes the gun particularly suitable for use in selected defensive positions, where the lack of mobility due to the rather heavy weight of the gun is not a serious handicap.

The Maxim gun weighs about 73 pounds including 7 quarts of water in the jacket.

The tripod for the Maxim weighs 80 pounds.

The Vickers gun weighs about 38 pounds including 7½ pints of water in the jacket.

The tripod for the Vickers weighs about 35 pounds.

The water in the jacket of the Vickers gun boils after 600 rounds of continuous fire, and after boiling has commenced, the water evaporates at the rate of about 1½ pints per 1,000 rounds.

These guns are noted for their reliability and freedom from jams and malfunctions.

PRINCIPLE OF OPERATION:

These guns are known as the recoil operated type. The barrel is movably attached to the rest of the mechanism, and carries on its muzzle end a disc which fits inside of a nozzle pierced with hole just large enough to allow the escape of the bullet. When the gun is fired the force of recoil, coupled with the reaction of the escaping gas between the nozzle and the barrel disc, drives the barrel about 1-inch to the rear. This movement suffices to operate the mechanism. The parts are then returned to their former position by the recoil spring.

MANIPULATION:

To load:

(a) Thread tag end of belt through feedbox from right to left.

(b) Grasp roller handle with right hand and swing it as far as it will go. Holding the handle in this position, sieze the tag end of belt with the left hand and pull it as far to the left as possible. Let go of roller handle.

(c) Repeat the motion described under (b).

To Fire: Grasp the handles with both hands, and with both thumbs press the safety catch down (Maxim) and press in the trigger. To stop firing, release the pressure.

To Unload: Work the roller handle through the complete motion by hand twice. With the fingers of the right hand reach under the edge of the feed box and press the extensions of the pawls. Draw out the belt. Then see that the gun is pointed in a safe direction, and press the trigger.

EXPLANATION OF THE ACTION OF THE MECHANISM IN LOADING AND UNLOADING.

When the belt is threaded through the feed box in loading, the first cartridge cannot be pulled into place, because the lock is in the way. When the roller handle is swung, the lock is moved and the cartridge can be pulled into place. When the lock is allowed to go forward, the carrier rises, gripping the cartridge in the belt. The next motion of the handle causes the lock to pull this cartridge out of the belt and place it in line with the chamber. The belt is then pulled, placing a fresh cartridge in position in the feedbox. When the handle is released, the lock goes forward, placing the first cartridge in the chamber. The carrier rises and engages the cartridge in the belt. The gun is then ready to fire. The reason the belt must be pulled each time is because the motion of the roller handle does not work the feedbox mechanism, a backward motion of the barrel, which only occurs on firing, being necessary for this.

For the same reason, the gun is unloaded when the roller handle is moved twice by hand. As the lock is drawn back, it pulls out the cartridge which is in the chamber and also the one which is in the belt, and as it goes forward it places them, respectively, in the ejection tube and in the chamber, and as the carrier rises it does not grasp a fresh cartridge because, as the barrel has not recoiled, none has been fed into place in the feedbox. The second motion of the roller handle withdraws the cartridge from the chamber and ejects it. The gun should then be pointed in a safe direction and snapped, in order to relieve the tension on the mainspring and also to make sure that no cartridge remains in the barrel.

To load so that only single shots will be fired.—The gun being fully loaded, work the roller handle through one complete motion. When the trigger is pulled one shot only will be fired, and each time thereafter that it is desired to fire a shot the roller handle must be worked by hand. This is sometimes useful in aiming exercises.

ACTION OF THE MECHANISM IN FIRING:

Suppose the gun to have just fired the first cartridge in the belt; the carrier will be gripping the second live cartridge in the feed block and the empty case, which has just been fired, in the chamber; the explosion will cause the recoiling portion to move backwards through a distance of about one inch, thereby extending the recoil spring.

This backward movement is due partly to recoil and partly to the effect of the muzzle attachment, which acts as follows:

The powder gases which escape from the muzzle after the exit of the bullet strike violently against the nozzle and rebound on to the front face of the barrel disc, driving it and the barrel, to which it is attached, backwards. The gases then escape into the air through the openings in the outer casing.

Action in the Feed Block.—The recess in the prolongation of the left recoil plate by means of the stud actuates the bottom lever of the feed block. The bottom lever acts on the top lever, which moves the slide and the top pawls to the right, to engage behind the cartridge held in place by the bottom pawls.

Rotation of the Crank.—During the first part of the backward movement, the lock is held tightly against the back of the barrel, and moves with it, supporting the cartridge in the chamber. This holding of the lock against the force of the explosion is accomplished by the fact that at the moment of firing the cross head and the crank are in a straight line, and the pressure in the chamber is transmitted through the lock, side levers, cross head, crank, and recoil plates, back to the barrel. As the backward motion of the parts continues, the cam surface of the roller handle strikes the roller, thereby rotating the crank and releasing the straight line locking of the parts which has, up to this time, existed. The rotation of the crank draws back the lock and causes the fusee to wind the fusee chain, thus further extending the recoil spring. The continued rolling of the crank brings the hump of the handle against the roller, and this, assisted by the recoil spring forces the whole of the recoiling portions forward again, with the exception of the lock, which continues its backward movement for a short distance before it joins in the forward movement. As the recoiling portions go forward, the recess in the prolongation of the left recoil plate actuates the bottom lever of the feed block. This bottom lever acts on the top lever which moves the slide and the top pawls to the left, the pawls thus bringing the third cartridge in the belt to a position against the cartridge and bullet stops, ready to be gripped by the carrier. The belt, as it moves to the left, slides over the bottom pawls, which are depressed as the cartridge passes over them, and rise behind the fourth cartridge, holding the belt in position and preventing it from sliding back after the second cartridge has been withdrawn by the carrier.

The Lock.—As the lock moves backwards the carrier withdraws the empty case from the chamber and a live cartridge from the belt in the feed block. The horns of the carrier move along the upper surface of the cams until the cartridge is clear of the belt. When the extractor arrives at the rear end of the cams it is forced down by the ramps in the cover, thus bringing the cartridge drawn from the feed block in line with the chamber, and, in case of the Maxim, bringing the empty case drawn from the chamber in line with the ejection tube. (In the Vickers the ejection tube is omitted. The empty cartridge falls off the carrier, or if it fails to fall off, is forced off when carrier rises, by the ejection seating on bottom of barrel casing.)

Cocking Action.—The rotation of the crank gives an upward motion to the connecting rod and side lever head, which latter, bearing on the tail of the tumbler, rotates it on its axis, and thus forces the firing pin to the rear. The long arm of the mainspring acts on the projection of the firing pin, while the short arm bears against the nose of the hand sear; consequently the withdrawal of the firing pin compresses the main spring by drawing the long arm towards the short arm. As the tumbler rotates, the nose of the hand sear is forced by the short arm of the main spring under the notch of the tumbler, and the continued motion of the tumbler forces the firing pin still farther back, until the notch of the safety sear (which is actuated by its spring) is forced into the notch of the firing pin and retains it. The firing pin is thus prevented from flying forward.

Action of the Recoil Spring. When the force of the explosion is expended, the action of the recoil spring comes into play, continuing the forward movement of the barrel and side plates and unwinding the fusee chain from the fusee. This gives the crank a rotary motion, which forces the cross head and side lever head downward, causing the lock to continue the forward movement, and place the live round in the chamber. The carrier is moved upwards by the side levers acting on the carrier. The bottom projection of the gib slides over the base of the live cartridge in the chamber, and the top projection of the gib slides over the base of the cartridge which has been automatically moved up into position in the feed block. The firing pin hole is thus brought opposite the cap. As soon as the carrier reaches its highest position, the carrier holding up spring engages in a groove in its sides to prevent the horns falling below and jamming on the front end of the cams in the breech casing at the commencement of the backward movement, when the side levers are released from the carrier. This, however, can only occur when there are no cartridges on the face of the carrier.

The further downward movement of the cross head and side lever head causes the lock to be forced slightly further forward, and the breech is then closed.

Firing Action.—(a) For the first shot. As the side lever is brought slightly beyond the horizontal, it raises the safety sear, thereby disengaging it from the firing pin, which then moves slightly forward until the notch of the tumbler engages the nose of the hand sear. If the safety catch is depressed and the thumb-piece on the trigger pressed, the trigger bar is drawn to the rear. As the trigger bar is drawn backwards, the front end of the slot engages and draws back with it the tail of the hand sear, thereby releasing the tumbler. The long arm of the lock spring then propels the firing pin on to the cap and the cartridge is exploded.

(b) For subsequent shots. The firer, by maintaining pressure on the thumb-piece, holds back the trigger bar. Therefore, each time the lock goes forward the front end of the slot holds back the tail of the hand sear before the lock is quite home. By this means the nose of the hand sear is prevented from engaging in the notch of the tumbler. When the lock is home, the side lever head raises the safety sear, thus permitting the long arm of the lock spring to carry the firing pin on to the cap, and the charge is exploded.

The releasing of the safety sear is so timed that the firing pin cannot be released until the lock is in the firing position.

On releasing the thumb-piece, the short arm of the lock spring forces the nose of the hand sear under the notch of the tumbler, so that, when the sear is depressed, the firing pin is unable to go forward.

ADJUSTMENT.

Adjustment of the Spring.—To insure the best operation of the gun, the recoil spring must be adjusted to the proper tension, which for the Maxim, is 7 to 9 lbs. with the barrel disc adjusted as described below. The tension is measured as shown in the photograph, using the spring balance furnished with the gun.

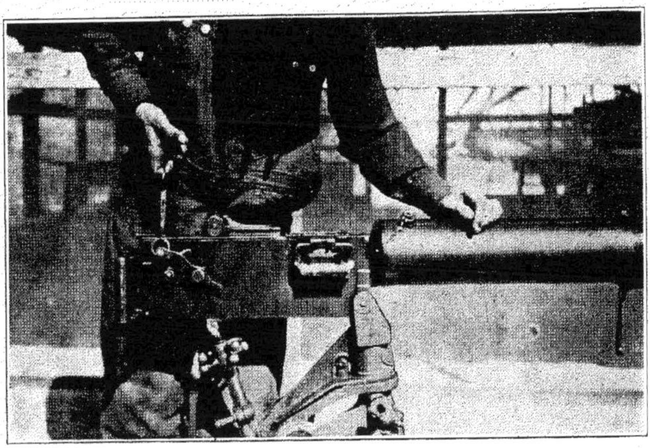

Testing tension of recoil spring

Adjustment of the Barrel Disc.—The operation of the Maxim is considerably affected by the position of the barrel disc, which should be adjusted by trial to give the best results. In the absence of an opportunity to determine the best setting experimentally, the following method will answer:

Screw the barrel disc on as far as it will go. Then unscrew it from 1½ to 2 complete turns and lock it in position by tightening barrel disc locking screw.

PACKING THE BARREL.

The packing at each end of the jacket where the barrel enters must be correct, for if it is too tight, excessive friction will interfere with the action of the mechanism, and if it is too loose, the water will run out. The method of packing is as follows:

For the rear bearing.—Wind the packing in the groove of the barrel, starting at the center and winding alternate strands on each side. As the groove fills up press the packing toward the center by inserting a knife blade between the side of the groove and the packing. The packing should be wound tightly until the groove is completely filled and the packing projects slightly above the surface of the barrel. The free end is then tucked down at the side and beneath several of the strands to prevent unwinding.

For the front end.—With the barrel in place and the follower removed, wind the packing around the barrel and continue pressing the coils in with a stiff wire until the seat is filled. Then screw in the follower. The proper amount of packing must be determined by trial.

After the packing is completed, the friction of the recoiling portions should be tested with the spring balance, as shown in the photograph. For the gun to work smoothly the friction should not be over 7 lbs. and it may be as low as 4 lbs.

Testing friction of recoiling parts. Note position of roller handle, also that recoil spring is removed during this test.

LOCK:

The lock should not be dismounted except by the company mechanic or a commissioned officer. Should the lock fail to function properly the spare one should be used until an opportunity for correcting the difficulty occurs.

Should it become necessary to dismount the lock, proceed as follows:

(1) Cock the lock.
(2) Force out inner side lever pin, using cupped end of handle block pin.
(3) Force out outer side lever pin, using square end of handle block pin.
(4) Take off side levers.
(5) Take off lifting levers.
(6) Slide off carrier.
(7) Push out tumbler pin, shake out tumbler.
(8) Keep fingers away from firing pin hole, and press up safety sear.
(9) Push out hand sear pin, keeping finger on bottom of lock to prevent main spring from flying out.
(10) Shake out hand sear and main spring.
(11) Lift up safety sear and shake out firing pin.
(12) Remove safety sear pin and safety sear.

To assemble:

(1) Put in safety sear and safety sear pin.
(2) Insert firing pin with notch for safety sear towards top of the lock.
(3) Insert tumbler, with tail towards the rear, and put in tumbler pin.
(4 Insert hand sear, with hump towards the rear, and put in hand sear pin.
(5) Press down tail of tumbler until you hear one click, then slide carrier on with horns towards the top.

(6) Put on the lifting levers.
(7) Put on side levers, and replace outer and inner side lever pins.
(8) Replace main spring as follows:
  (a) Press up side lever head.
  (b) Press up tail of tumbler.
  (c) Pull back hand sear.
  (d) Place main spring in front of hand sear with long arm to the front and force down spring with thumb until it snaps home with a click.

ADJUSTMENT OF THE CROSS HEAD:

Ruptured Cartridge Cases.—If the cross-head is not long enough, the lock will not be held tightly against the face of the cartridge, and bulged or ruptured cartridge cases will result, as described in the introduction to part I.

When this difficulty occurs during firing, take off the lock, place one of the thin adjusting washers on the crosshead *in front of* the adjusting nut, replace lock, and continue firing. At the first opportunity, take off the washer, place it *behind* adjusting nut, and screw up adjusting nut.

If the cross-head is too long, the lock will press so tightly against the head of the cartridge that it cannot go home. In this case remove one or more of the thin washers behind the adjusting nut.

To test the length of the Cross Head.—
(1) Take off the lock.
(2) Place over the end of cross head a .005 washer.
(3) Replace lock.
(4) Take off recoil spring.
(5) Place a new dummy over the firing pin hole by sliding it on to the carrier.
(6) Work the lock to cause the dummy to enter the chamber.

(7) Be sure the recoiling portions are home.

(8) Press down the roller handle.

(9) If the cross-head was of the right length before the extra washer was put on, a slight check will be felt; and in this case the washer should be removed. If the slight check is not felt, it shows that the cross-head was too short. In this case, place the washer *behind* the adjusting nut, place another washer on the front end, and test as before.

LOADING MACHINE:

This is shown in the photograph. It should be adjusted so that the cartridges are pushed in until the points of the bullets are even with the ends of the brass tags on the belt.

Loading machine

In operating the loading machine, the handle should be held lightly and not turned too fast, and a momentary pause should be made at the point where the needle enters the pocket, so that if it does not enter properly it will not pierce and spoil the belt.

POINTS TO BE OBSERVED BEFORE, DURING, AND AFTER FIRING:

Before Firing.—

(a) Dismount, clean and oil, and assemble gun.
(b) See that packing is all right.
(c) Test friction of recoiling portions.
(d) See that barrel disc is properly adjusted.
(e) Test crosshead for length.
(f) Weigh and adjust recoil spring.
(g) See that barrel casing and water boxes are full.
(h) See that spare parts are in place and oil can full.
(i) See that belts are properly loaded and in good condition.

During a temporary cessation of fire.—

(a) Refill the water jacket.
(b) Oil the parts.
(c) Replace a partly used belt with a full one.
(d) Attend to any necessary adjustments or repairs.

After Firing.—

(a) Unload.
(b) Empty water jacket.
(c) Dismount gun.
(d) Clean and oil bore once a day for ten days.
(e) Reload used belts.
(f) Fill oil cans and water boxes.

JAMS AND MALFUNCTIONS.—

The following table lists some of the things which may cause trouble:

I. *Improper Manipulation:*
- Improper adjustment of barrel disc.
- Improper adjustment of spring.
- Lack of oil.
- Excessive friction, due to tight packing.
- Improperly filled belts.
- Improper loading.
- Trigger bar wrongly assembled.

II. *Broken Parts:*
- Gib.
- Gib spring.
- Lifting levers.
- Firing pin.
- Main spring.
- Fusee.
- Fusee links.

III. *Worn or Deformed Parts:*
- Enlarged chamber.
- Weak pawl springs.
- Tight belts (new or wet).
- Old belts, torn or with loose pockets.
- Bent brass tags on belts.
- Excessive headspace.

IV. *Defective Ammunition:*
- Misfire.
- Battered ammunition.
- Loose primers.
- Loose bullets.

When a stoppage occurs, the probable cause and the immediate action to be taken are best determined by noting the position of the roller handle. It may stop in four positions, which are as follows:

FIRST POSITION:

See photograph. Lock back more than a cartridge length.

Probable Cause.—

(a) Recoil spring too weak.
(b) Excessive friction, due to lack of oil, tight packing, etc.

(c) Tight pockets in belt (new or wet).

Immediate Action.—Complete the motion of roller handle by hand if possible. If the trouble recurs, loosen the recoil spring by several turns of the adjuster, and re-oil the parts.

Maxim gun, first position stoppage

Method of setting up for purposes of instruction.—

(1) Half load.

(2) Pull roller handle until the cartridge is just withdrawn from belt.

(3) Pull belt.

SECOND POSITION:

Lock back less the length of a cartridge. See photograph.

Probable Cause.—

(a) Damaged cartridge.

(b) Separated cartridge case, due to excess headspace.

Immediate Action.—Lift up cover, withdraw lock, examine cartridge. If damaged, remove it, reload and continue firing.

Maxim gun, second position stoppage

If cartridge is a good one, remove separated case from chamber with defective cartridge extractor. Place two adjusting washers in *front* of adjusting nut, replace lock, reload, and continue firing. After the engagement, place the washers *behind* the adjusting nut.

THIRD POSITION:

Lock almost home but carrier not entirely up. See photograph.

Probable Cause.—

(a) Recoil spring too weak.
(b) Foreign substances or excessive friction in the lock.
(c) Ammunition box wrongly placed, belt out of line with feed box.
(d) Belt being pulled.
(e) Badly filled belt, cartridges not home.
(f) Loose pockets in belt.

Third position stoppage. Note that roller handle lacks a half an inch of being onto roller

(g) Bent brass tags on belt.
(h) Broken firing pin, point jammed in lock.
(i) Broken gib.
(j) Broken gib spring.

Immediate Action.—

(1) Strike the handle down, continue firing.

(2) If (1) fails, lift up the handle slightly. (If the handle cannot be lifted, raise the cover, and rock the handle while the horns of carrier are forced down with a screwdriver or other instrument.) Pull belt to the left, let go of lock and continue firing. If first cartridge cannot be pulled into place in feed box, lift up feed box, press pawls, draw out belt and straighten up cartridges or bent brass tags, replace feed box, load, and fire.

(3) If (2) fails, change lock, as a broken gib, broken gib spring, point of a broken firing pin, or some other foreign substance may be binding the carrier.

To Set up this Stoppage.—

(a) To simulate a weak recoil spring—

(1) Half load.

(2) Then repeat, but check roller handle in third position.

(b) To simulate a slight fault in feed—

(1) Half load.

(2) Draw back the lock by swinging the roller handle.

(c) Pull next cartridge halfway over face of feed block.

(d) Let lock go home until it is checked by this cartridge.

(e) Push down on roller handle to be sure cartridge has not slipped.

To simulate a bad fault in feed.

(a) Pull fourth cartridge partly out of belt.

(b) Half load.

(c) Pull roller handle to a vertical position.

(d) Grasp roller handle by knob and tail, and draw recoiling portions back.

(e) Pull belt.

(f) Let recoiling portions go forward.

(g) Swing handle forward (Maxim), and let it go back again.

FOURTH POSITION:

Lock fully home. See photograph.

Probable Cause.—

(a) Misfire.

(b) Broken firing pin.

(c) Broken main spring.

(d) No cartridge in chamber, due to empty space in belt, or improper loading.

Maxim gun, fourth position stoppage

Immediate Action.—

(1) Half load, continue the fire.

(2) If this fails, unload, change lock, load, and fire.

Method of Setting up.—

(a) Misfire.

(1) Put a dummy cartridge in belt as 4th cartridge.

(2) Load.

(b) Broken firing pin or mainspring:

(1) Put dummies in belt at 2nd, 3rd, and, 4th, cartridges.

(2) Load.

(c) No cartridge in chamber:

(1) Half load.

(2) Work crank handle through its full motion without pulling belt.

SCHEDULE OF INSTRUCTION:

The following schedule of instruction is intended to familiarize the members of the organization with the mechanism of the gun. It should be supplemented by target practice and instruction in minor tactics.

PROGRAM OF INSTRUCTION FOR COMPANIES ARMED WITH THE MAXIM MACHINE GUN, MODEL OF 1904, or the VICKER'S MACHINE GUN, MODEL OF 1914

1. The course will consist of one week spent in study of the mechanism of the piece, and one week spent in exercises, which involve firing. The hours of instruction are from 8 to 11:00 A. M. and from 1:30 to 4:00 P. M.

2. Program for first week is as follows:

MONDAY FORENOON:

Lecture on the history and development of machine guns and their mechanism.

Mechanism. Principles of gun explained briefly by the instructor.

Work for Class. Disassembling and assembling, and thorough nomenclature.

MONDAY AFTERNOON:

Instruction in nomeclature continued by company officers. When each member of class can pick out every part mentioned in the alphabetical list of components of the rifle, and can name any part of the gun picked up at random by the instructor he will be considered sufficiently instructed in nomenclature.

TUESDAY FORENOON:

Lecture on characteristics which affect the tactical employment of Machine Guns.

Mechanism. Instruction in proper method of assembling and disassembling the gun in which the precautions necessary to prevent abuse of the gun in taking it down and setting it up are carefully explained. Explanation of Loading Machines.

Work by Class. Nomenclature of parts of loading machine. One machine will be disassembled for this purpose.

Nomenclature of parts of tripod. One tripod will be disassembled for this purpose.

TUESDAY AFTERNOON:

Assembling and disassembling gun blindfolded, to simulate night work.

WEDNESDAY FORENOON:

Lecture.—Mechanism of piece and function of each part.

Mechanism.—Instruction of class by officers and noncommissioned officers of the company in the mechanism of the piece, two guns to be used by each class. One gun will be assembled with recoil spring left out and placed on tripod, and the other completely disassembled (except water jacket) will be placed on the table. As the movement of the gun in firing is explained it will be illustrated as far as possible on the assembled gun and the student will be required to show the action by moving the pieces of disassembled gun in such a manner as to simulate the movement of the corresponding part in the operation of the gun. Dummy cartridges are to be used in this exercise.

WEDNESDAY AFTERNOON:

Functioning of piece as above, continued.

THURSDAY FORENOON:

Lecture—Jams, malfunctions, and stoppages. Their prevention. Care and preservation of the gun.

Work for Class. Replacing of parts. Effort should be made to learn the most expeditious method of replacing the following parts.—

Lock

Feed box

Recoil spring

Fusee and Links

Instruction in the method of packing barrel.

Instruction for officers, noncommissioned officers and mechanic in assembling and disassembling lock and feed box.

THURSDAY AFTERNOON:

Continue work of morning and review prior instruction.

FRIDAY FORENOON:

Lecture. Inspection of gun to verify its readiness for action.

Mechanism. After the lecture each gun will be gone over in the manner indicated and any faults detected will be immediately remedied.

Work for Class. Drill in loading and unloading the gun. Effort will be made to develop greatest possible team work among members of gun squad.

FRIDAY AFTERNOON:

Reloading of all belts and adjustment of loading machine.

SATURDAY FORENOON:

Lecture on the indication and recognition of targets. Practice in applying principles embodied in lecture, employing groups of targets concealed in the terrain near the school.

SECOND WEEK

MONDAY FORENOON:

Adjustment of each piece by actual firing and recording of all adjustment data for each gun.

MONDAY AFTERNOON:

Explanation of use of sights, setting, adjustments, etc.

TUESDAY FORENOON:

This period will be spent in familiarizing the entire company with the actual firing of the gun. This will be done by allowing each man to fire 20 rounds at a designated target.

TUESDAY AFTERNOON:

Loading practice. Each member of the company will be required to act as loader and afterwards as gunner loading and firing 5 belts with 5 cartridges in each end.

WEDNESDAY FORENOON:

Loading practice continued.

WEDNESDAY AFTERNOON:

Exercise in malfunctions, stoppages and jams. Each member of each gun squad will be required to act as gunner and fire a gun which has been previously put out of order in such a manner as to produce a predetermined malfunction, stoppage or jam, the nature of which will be unknown to the man firing the gun. He will be required to get the gun in action and fire a given number of shots, for some of which defective ammunition will be used, in the shortest possible time.

THURSDAY FORENOON AND AFTERNOON:

Malfunctions, stoppages and jams continued.

FRIDAY FORENOON:

Shot grouping. For this firing targets 1, 2, 3, and 4 will be used. Exercises in traversing and searching.

FRIDAY AFTERNOON:

Practice in use of night firing box and in indirect laying.

Practice to determine how many aimed shots each squad can fire in one minute.

SATURDAY FORENOON:

This period will be devoted to a firing problem at designated silhouette targets concealed in positions adjacent to the instruction range. Each squad will be given 2 boxes of ammunition (500 rounds). The company officers and scouts will be allowed to reconnoiter the terrain for the purpose of locating the hostile line, after which the company will be required to come up to the firing point from cover, go into action, and fire all ammunition alloted for the problem. After this exercise the hits on the targets will be counted and the instruction of the company will end.

CHAPTER IV

THE COLT AUTOMATIC GUN

This description is written for the 1914 model, but will serve equally well for older models, as the mechanism is practically the same, the main difference being that the later model has an interchangeable barrel.

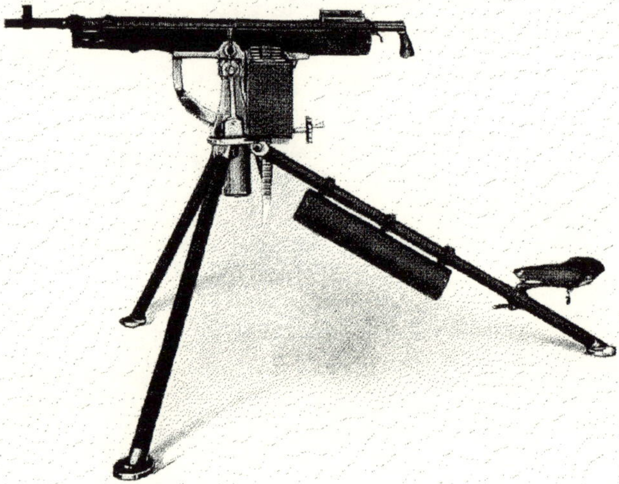

Gun mounted on tripod, cartridge box attached (Left side)

The gun, which is shown in the photograph, is of the air-cooled, gas operated type, fed from a belt and mounted on a tripod.

A distinguishing feature of the gun is the fact there are no adjustments to be made by the soldier, so that one of these guns in good condition can be operated by anyone who knows how to load it.

This gun is adjusted at the factory to shoot at a rate of slightly greater than 400 shots per minute.

The weight of the gun is about 35 lbs. The weight of tripod and mount is about 56½ lbs.

The cartridges are fed from belts holding 250 rounds each. These belts are filled by a special belt loading machine supplied with the gun. The filled belts are carried in special boxes which can be quickly attached to the mount as shown in the photograph. After the box is attached to the mount it moves with the gun, so that the belt cannot get out of line with the feeding mechanism.

PRINCIPLE OF OPERATION:

In the bottom of the barrel about 8 inches from the muzzle there is a vent, and pivoted in the casing under the barrel there is a lever, called the gas lever, which carries at its free end a cylindrical plug, called the piston, which rests against the vent. These parts can be seen in the sectional view of the gun herewith. When the gun is at rest the gas lever is held in this position by two retracting springs to which it is connected by a series of links. When the gun is fired a part of the powder gas escapes through the vent and impinges against the piston, causing the gas lever to rotate downward and backward against the action of the retracting springs. This movement of the gas lever through its connections unlocks and pushes back the bolt, drawing out and ejecting the empty shell and cocking the hammer. At the same time the cartridge extractor (82) draws a fresh cartridge out of the belt onto the carrier (21). When the energy which caused the backward motion of the gas lever is expended, the retracting springs come into play and rotate the gas lever forward and upward again. This return motion of the gas lever causes the front end of carrier to rise, bringing the cartridge on carrier in front of bolt. The bolt is then carried forward placing the new cartridge in the chamber. At the same time the feed wheel (61) rotates placing a fresh cartridge in position to be engaged by the cartridge extractor, which is now moving forward.

During the final motion of gas lever, the bolt pin (14) in the cam in rib of bolt cants the rear end of the bolt down, causing it to lock behind the shoulders in the receiver. The trip then releases the sear, and if the trigger is also being held back, the hammer is released and flies forward striking the firing pin and the cartridge is fired, causing a repetition of the cycle.

MANIPULATION

To Load:

(a) Push the brass tip of the belt through the opening above the box and pull it out on the other side of the gun as far as it will go.

(b) *Let go of the belt.* (Never pull the belt while the gas lever is being operated.)

(c) Swing gas lever downward and to the rear *until it strikes the bottom plate of the gun.*

(d) Release gas lever, which will resume its normal position.

The gun is now ready to fire.

To Fire.—Hold back the trigger.

To hold the gun ready for firing.—Push the safety latch on the right of the gun from "Fire" to "Safe." This will prevent the trigger from releasing the hammer.

Never leave the gun loaded after it has been firing. The heat of the barrel may explode the cartridge.

To unload.—Press forward the knurled head of the throw off on the right hand side of the gun near the belt exit, and draw the loaded belt out from the left. Then operate the gas lever once, as in loading, to eject the cartridge that still remains in the chamber.

To Remove Interchangeable Barrel.—Take hold of gas lever pin and throw the gas lever rearward until it strikes the bottom plate, and hold it in this position. Withdraw handle lock (2) and insert small end of it in small hole in rear end of gun on right side, pushing handle lock in as far as it will go. This causes the bolt pin (14) to lock the slide to receiver and to hold the gas lever back. Bolt pin will then project from the left hand side. Leave gun in this position. Replace handle lock.

Attached to gas cylinder (27) will be found barrel lock (A); place small end of combination wrench (C) under point of barrel lock at A, and press down and out until it clears barrel. Engage combination wrench in slot (B) in under side of barrel just forward of gas cylinder and unscrew; then pull the barrel forward and out.

To Replace Interchangeable Barrel.—Push the barrel in as far as it will go, and turn it a few times by hand to engage the threads; complete this operation with the combination wrench until the lines on top of barrel and receiver coincide. Swing the barrel lock up into slot. Release the gas lever by pushing in the bolt pin from the left hand side as far as it will go, with handle lock. Replace handle lock. Do not attempt to remove or replace the barrel until the gas lever has been thrown to rear, and locked in that position.

To Dismount the Breech Mechanism.—

First: Take hold of the gas lever pin and throw the gas lever rearward until it strikes the bottom plate. Then release the gas lever, when it will fly forward to its original position. This movement cocks the hammer so that it can be removed with the handle.

Second: On the right hand side of the gun, forward of the trigger, is the handle lock, which is a pin with a small projecting lever resting in a horizontal position. Turn this lever upward and backward as far as possible. Draw the handle lock out to the right. This releases the handle, which can then be moved rearward. The handle contains the mainspring, hammer, sear, sear spring, trigger and trigger spring. To remove the hammer and mainspring from the handle, pull back the trigger, release the sear by pulling the nose down, when the hammer and mainspring will fly out, as the sear holds them in place.

Third: To remove the bolt throw the gas lever rearward as far as possible, and while holding it in that position insert the small end of the handle lock as far as it will go. This removes the bolt pin from the bolt. Withdraw the handle lock, but leave the bolt pin in the position in which it now is, i. e., projecting from the left side. The bolt is now free to be removed from the rear of the gun.

Fourth: The extractor and firing pin can be removed from the bolt by pushing out the small pins that hold them in place.

To Assemble the Breech Mechanism.—

First: Insert the bolt and push forward as far as it will go.

Second: Place the hand on the gas lever holding it up against the bottom plate.

Third: Push in the bolt pin from left hand side as far as it will go.

Fourth: Then release the gas lever.

Fifth: Replace the handle, locking it in position with handle lock.

To Completely Dismount the Gun.—Lay the gun on its right side, muzzle to the left (upon 2 pieces of wood if possible.) Turn the side plate lock screw (56) to left 1/4 turn, or enough to have cut in it clear groove in side plate screws (55 and 57). Unscrew side plate screws (55 and 57) and remove them. Place the right hand on the rear of left hand side place (90) and gently press downward, and with the left hand raise front of plate. Turn the gun over and remove the right hand plate (89) in the same manner. Turn the gun so as to bring the bottom plate (91) upward. Raise gas lever slightly with left hand, 3 or 4 inches, and lift the bottom plate out with the right hand. Raise the gas lever again and pull out the slide pin (87). Push the slide (86) to the rear until it remains. Remove the retracting spring tubes (39 and 40) by pulling the nearest tube toward you and pushing the other away from you. Pull or push out gas lever bracket pin (34), then lift the gas lever and its connections (31 and 45) out of the bracket (33). Push carrier pin (22) out with the pointed end of screw driver and lift carrier (21) out of receiver. Unscrew the locking screw and the belt guide screw (6) in the belt guide (59). Lift belt guide (59) out. If the belt guide should stick push the point of the screw driver under the slide at rear of belt guide and gently pry upward. Remove the handle lock (2) and pull the handle (1) out to rear. Push the slide (86) to the rear and insert the end of handle lock (2) in the small hole in rear of the receiver in the side nearest to you, and push out the bolt pin (14). Pull the slide (86) to the front and out and then push bolt to rear with the finger. Remove safety (58) and trip (85).

To Assemble.—Hold gun with the slot in receiver upward. Insert the bolt, front end first. Replace the slide, pushing to the rear as far as it will go. Replace the bolt pin, pushing it in as far as it will go by using end of the handle lock. Replace the belt guide and screw, having locking screw engage groove in belt guide screw. Replace the carrier having flat side or bottom up and carrier dog (23) down. Insert pin (22). Replace the gas lever having piston (35) drop into hole in gas cylinder (27). Adjust the gas lever until large pin holes in the gas lever and the gas lever bracket are opposite each other, then insert the gas lever bracket pin (34). Next place the retracting spring tubes in position, having slot in followers (41 and 42) align with long slot in retracting spring tubes. Force follower over projecting end of retracting connection (45) by pressing follower end inward, and at the same time slightly compressing retracting springs (37 and 38) until pilot at end of the tube clears the bracket (33). Press the tube inward at the bracket end and into the hole for pilot. Place the bottom plate in position by putting tongue at rear end in receiver first and flush with rear end of receiver. Place safety (58) in position. The right hand side plate (89) should be next placed in position, taking the PRECAUTION to have slide (86) pushed to the rear as far as it will go and the feed lever (66) on plate 89 carried to rear as far as it will go, so that feed lever (66) will lie between the projections on the slide (86) when the plate (89) is in position. To place the plate (89) in position have the dowel at rear of the plate enter hole for it, in the receiver, at the same time having the groove near the edge of the plate engage the rib on the bottom plate and being sure that the gas lever bracket pin (34) enters hole in front end of plate (89). Turn gun over to rest on plate (89), draw the slide (86) forward, having slot at forward end of the slide engage end of gas lever connection (31). Hold the gun at rear end with right hand to prevent it from moving. Take hold of gas lever pin with the left hand and swing gas lever outward until the slide pin (87) can be pushed into the hole in the slide (86) through gas lever connection (31), having flatted end of the slide pin (87) enter slot in side plate. Place trip (85) in holes and slot in receiver. Place left hand side plate (90) in position by the same method followed with the right hand plate. Place side plate screws (55 and 57) into holes for them in the front and rear of side plates. If screws do not enter freely see that bottom plate is flush with rear of receiver. If it is not rap the bottom plate gently on front end until it is flush with rear end of receiver. Screw the plates together and lock with the locking screws (56) making a quarter turn of these screws to engage the grooves in the side plate screws. Push the handle (1) into the hole in rear of receiver and lock it with handle lock (2). The handle lock is placed in position by inserting large end into the hole with the stem held upward and then turning it to the front and downward to have the projection gauge locking cut in the side plate, and so that the handle lock stop (3) will spring over the stem. Place the gun in the mount and run a few cartridges (preferably dummies) in a belt through gun to see that the mechanism functions properly. Keep the hand away from the trigger during this operation, unless you use dummies, and be sure to work the gas lever until all the cartridges are out of the gun. To obtain more leverage with large screw driver insert the blade of the small screw driver into hole in large screw driver and use it as a lever.

BELT LOADING MACHINE:

This machine is shown in the photograph.

To Fill a Belt with Cartridges.—Fasten the machine to a table or bench and turn the crank (10) to the right until it is straight down.

Release the tension-spring hook (28) and raise the upper feed wheel (18) as far as it will go.

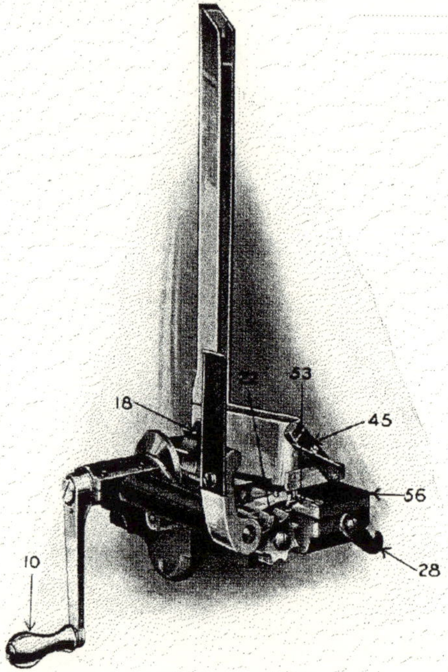

Belt-loading machine in position awaiting insertion of belt

Turn the belt guide cover (56) to the right far enough to admit the belt into the belt guide (53) and raise the upper needle bar (45) as far as it will go.

There is a black thread running on the side of the belt which indicates the side for the entrance of the cartridge. Put two cartridges by hand into the two loops of the belt nearest the end with the brass tip, and place the belt in the machine with the first cartridge resting in the top groove of the lower feed wheel (22), and the belt passing out at the back through the belt guide.

Return the belt guide cover to place over the belt (being careful to see that the belt is free to pass under it), and lower the needle bar.

Turn the upper feed wheel down upon the belt and secure the tension spring under the hook.

Fill the feed guide with cartridges by stripping them from the clips.

Turn the crank to the right and the cartridges will be fed into the belt ready for use in the gun.

Place a feed box ready to receive the filled belt and at such a height that not more than two feet of filled belt will be suspended from the feed wheels of the machine.

Precautions.—Before beginning to use the machine, see that it is well oiled and that the needles are properly set. The needles should be so placed as to have their points even (in same vertical line), and about 1/100 of an inch apart.

In case of a miss in charging the belt, stop and open up the machine and remove the belt. Turn the crank to the right until straight down, as in starting, and replace the belt in machine with the *Next to the Last* cartridge in the top groove of the lower feed wheel. Close the machine and proceed as before.

MALFUNCTIONS, STOPPAGES, AND JAMS:

General Rule.—In case of a stoppage, look at the gas lever.

(a) If it is closed, work it once by hand and attempt to continue the firing before taking any other action.

(b) If it is partly open, push it *backward* until it strikes the bottom plate. *Never push it forward.* Then let go and if it goes forward by its own force, continue firing. If it fails to go forward, look for some of the troubles mentioned below.

Position of Gas Lever.—In case of a malfunction the gas lever stops in one of three characteristic positions, and an observation of the position in which it stops is a great help in indicating the probable nature of the trouble. These positions are shown in the photographs, and are as follows:

First characteristic position of gas lever

First Position.—Gas Lever closed.

Probable cause of stoppage.

(a) Misfire.

(b) Failure of sear to hold back the hammer.

(c) Empty space in the belt.

Immediate Action.

(1) Wait a moment to be sure it is not a hang fire. Then work the gas lever by hand once and continue firing.

(2) If the trouble recurs after one shot, the trouble is probably due to (b). Take out handle and put in a new sear spring, or if sear and spring are all right, examine the hammer, and if rim is worn, use a spare hammer.

(3) If no cartridge is ejected when the gas lever is worked, the trouble is probably due to (c). Pull belt to the right as far as it will go. *Let go of belt.* Work gas lever once and continue firing.

Second Position.—Gas lever about half way back.

Probable cause.

(a) Damaged cartridge.

(b) Ruptured cartridge, caused by too much head-space.

(c) Belt improperly loaded, cartridge not pushed far enough in.

Immediate Action.

Pull gas lever back and let it go forward. If the trouble was due to (a) firing can be resumed. If after the gas lever is pulled back, it fails to go forward and stops at the same place, look in ejection opening and see if a fresh cartridge is partly entered in chamber.

(1) If so, pull back the gas lever, hold it back and remove ruptured case with defective cartridge extractor. Let gas lever go forward and continue firing.

(2) If no cartridge is being fed into chamber, look at cartridge just entering feed slot at the left side of gun. If it is partly out of belt and jammed against side plate, hold back gas lever, push cartridge into belt, let gas lever go, and continue firing.

Second characteristic position of gas lever

Third Position.—Gas Lever nearly all the way back.

Probable cause:

(a) Failure to extract. Live shell pushing against empty one in chamber.

Third characteristic position of gas lever. Note that gas lever is all the way back.

(b) Cartridge only partly pulled out of belt by cartridge extractor, preventing upward motion of carrier.

Immediate Action.

(1) Look into ejection opening, and if there is an empty case in the chamber, hold back gas lever, letting live cartridge descend onto carrier. Then with a screwdriver or the point of a bullet pry the empty case out of the chamber and remove it. Let go of gas lever and continue firing. If the trouble recurs, use the spare bolt until an opportunity can be found to change the extractor and extractor spring.

(2) If there is no cartridge in the chamber, the trouble is probably due to (b). Hold back the gas lever and with a screw driver or knife blade, reach through the slot in the right hand side of the gun below the ejection opening and draw the cartridge back onto the carrier. Let go of gas lever and continue firing.

GENERAL PRECAUTION

When the gun is very hot, as after firing, a cartridge, if it is left in the chamber more than six or seven seconds, may explode from the heat. In case of a stoppage of any kind, if firing cannot be resumed in a few seconds, *withdraw the belt and unload the gun.* Remember that the gun may be discharged by the heat, regardless of whether or not the safety is on. KEEP AWAY FROM THE MUZZLE OF THE GUN.

PROGRAMME OF INSTRUCTION:

The instruction outlined in this programme is intended to familiarize the personnel of the organization with the mechanism of the gun, and it should be supplemented by instructions in minor tactics and by target practice.

PROGRAM OF INSTRUCTION FOR COMPANIES ARMED WITH THE COLT AUTOMATIC MACHINE GUN

1. The course will consist of one week spent in study of the mechanism of the piece, and one week spent in exercises, which involve firing. The hours of instruction are from 8 to 11 A. M. and from 1:30 to 4:00 P. M.

2. Program for first week is as follows:

MONDAY FORENOON:

Lecture on the history and development of Machine Guns and their mechanism.

Mechanism. Principles of gun explained briefly by the instructor.

Work for Class. Disassembling and assembling and thorough nomenclature.

MONDAY AFTERNOON:

Instruction in nomenclature continued by company officers. When each member of class can pick out every part mentioned in the alphabetical list of components of the rifle, and can name any part of the gun picked up at random by the instructor, he will be considered sufficiently instructed in nomenclature.

TUESDAY FORENOON:

Lecture on characteristics which affect the tactical employment of Machine Guns.

Mechanism. Instruction in proper method of assembling and disassembling the gun in which the precautions necessary to prevent abuse of the gun in taking it down and setting it up are carefully explained. Explanation of Loading Machines.

Work by Class. Nomenclature of parts of loading machine. One machine will be disassembled for this purpose.

Nomenclature of parts of tripod. One tripod will be disassembled for this purpose.

TUESDAY AFTERNOON:

Assembling and disassembling gun blindfolded, to simulate night work.

WEDNESDAY FORENOON:

Lecture. Mechanism of piece and function of each part.

Mechanism. Instruction of class by officers and Noncommissioned officers of the company in the mechanism of the piece, two guns to be used by each class. One gun will be assembled and placed on tripod, and the other completely disassembled will be placed on the table. As the movement of the gun in firing is explained it will be illustrated as far as possible on the assembled gun and the student will be required to show the action by moving the pieces of the disassembled gun in such a manner as to simulate the movement of the corresponding part in the operation of the gun. Dummy cartridges will be used in this instruction.

WEDNESDAY AFTERNOON:

Functioning of piece as above, continued.

THURSDAY FORENOON:

Lecture, Jams, Malfunctions and Stoppages. Their prevention. Care and preservation of the gun.

Work for Class. Replacing of parts. Effort should be made to learn the most expeditious method of replacing the following parts:

Barrel	Mainspring
Bolt	Hammer

Instruction for officers, noncommissioned officers, and mechanic in replacing firing pin, shell extractor, shell extractor spring, cartridge extractor, trigger spring, and sear spring.

THURSDAY AFTERNOON:

Continue work of morning and review prior instruction.

FRIDAY FORENOON:

Lecture. Inspection of gun to verify its readiness for action.

Mechanism. After the lecture each gun will be gone over in manner indicated and any faults detected will be immediately remedied.

Work for Class. Drill in loading and unloading the gun. Effort will be made to develop greatest possible teamwork among members of gun squad.

FRIDAY AFTERNOON:

Reloading of all belts and adjustment of loading machines.

SATURDAY FORENOON:

Lecture on the indication and recognition of targets. Practice in applying principles embodied in lecture, employing groups of targets concealed in the terrain near the school.

SECOND WEEK

MONDAY FORENOON:

Test of each piece by actual firing and correction of all faults that may develop.

MONDAY AFTERNOON:

Explanation of use of sights, setting, adjustment, etc.

TUESDAY FORENOON:

This period will be spent in familiarizing the entire company with the actual firing of the gun. This will be done by allowing each man to fire 20 rounds at a designated target.

TUESDAY AFTERNOON:

Loading practice. Each member of the company will be required to act as loader and afterwards as gunner loading and firing 5 belts with 5 cartridges in each end.

WEDNESDAY FORENOON:

Loading practice continued.

WEDNESDAY AFTERNOON:

Exercise in malfunctions, stoppages and jams. Each member of each gun squad will be required to act as gunner and fire a gun which has been previously put out of order in such a manner as to produce a predetermined malfunction, stoppage or jam, the nature of which will be unknown to the man firing the gun. He will be required to get the gun in action and fire a given number of shots, for some of which defective ammunition will be used, in the shortest possible time.

THURSDAY FORENOON AND AFTERNOON:

Malfunctions, Stoppages, and Jams, continued.

FRIDAY FORENOON:

Shot grouping. For this firing targets 1, 2, 3, and 4 will be used. Exercises in traversing and searching.

FRIDAY AFTERNOON:

Practice in use of night firing box and in indirect laying. Practice to determine how many aimed shots each squad can fire in one minute.

SATURDAY FORENOON:

This period will be devoted to a firing problem at designated silhouette targets concealed in positions adjacent to the instruction range. Each squad will be given 2 boxes of ammunition (500 rounds). The company officers and scouts will be allowed to reconnoiter the terrain for the purpose of locating the hostile line, after which the company will be required to come up to the firing point from cover, go into action, and fire all ammunition alloted for the problem. After this exercise the hits on the targets will be counted and the instruction of the company will end.

PART II

THE PRACTICAL HANDLING OF MACHINE GUN FIRE

CHAPTER V

TARGET DESIGNATION

In the designation and description of targets, the system used is practically the same for machine guns as for the shoulder rifle. It is, however, made much easier by the difference in the number of men whose attention it is necessary to direct. It is only essential for the target to be picked up by one individual, the man who is firing the gun. If necessary, the gun may first be directed on the target by an officer, and afterwards the gunners may locate the target by looking over the sights.

In designating targets, no time should be lost in a verbose description of a target that stands out so plainly that it is self evident what objective is meant. Under such circumstances the system of designating the target would be as follows:

Method	Example
(a) Announce target.	Target, Hostile Machine Gun.
(b) Announce range.	Range, 800 yards.

In trench warfare, where the gun is already in position for firing, it is better to announce range first, so that the gunner can at once set his sight. In all other cases it saves time to announce the target first, for the gunner can be observing the target while the gun is being brought into action, but he cannot set the sight until the gun has been set up.

Reference Points. In those cases where the target is an indistinct object, and is difficult to pick up without the aid of field glasses, it becomes necessary to select some prominent object in the vicinity which can be used as a reference point to facilitate the description of the target. Any object which will quickly command attention in that particular front will do.

There are a number of methods by which indistinct targets may be described, but it must be remembered that elaborate or lengthy explanations are difficult to grasp and should not be used unless it is necessary.

The methods which have been found best by experience are given below. The method which is simplest and clearest for the particular objective and terrain with which one is confronted should be used.

The examples given in connection with the following descriptions are intended for illustration of target designation only, and do not necessarily illustrate the correct tactical principles of machine gun fire.

Horizontal Clock face System. This method is suitable for targets which are plainly visible.

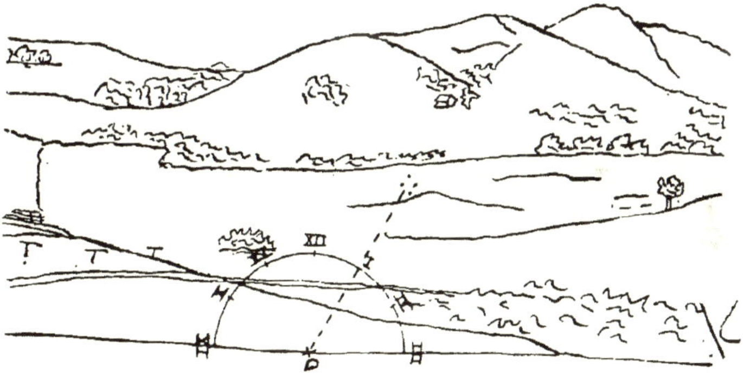

METHOD	EXAMPLE
(a) Announce direction.	At 1 o'clock.
(b) Announce target.	Target, a patrol.
(c) Announce range.	Range, 900 yards.

1. All look along a line pointing towards 1 o'clock of an imaginary horizontal clock dial, whose center is at the firing point, and whose 12 o'clock line is perpendicular to the firing line.

Vertical Clock Face System. This system is suitable for indistinct targets.

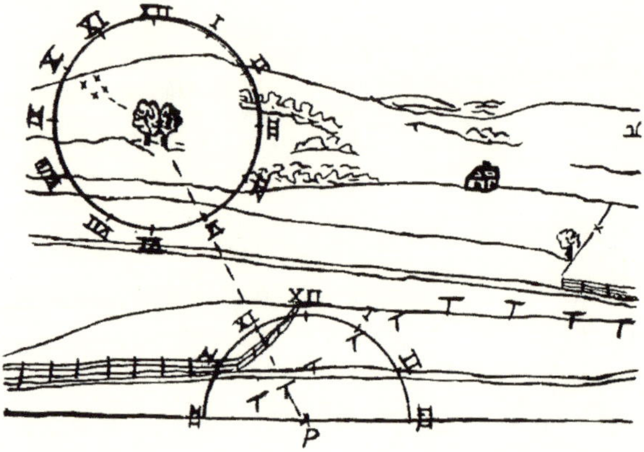

	METHOD	EXAMPLE
(a)	Announce general direction of reference point.	"Reference point to our left front" or "at 11 o'clock."
(b)	Select most prominent object on indicated zone.	"Two trees."
(c)	Describe position of target with respect to reference point.	"At 10 o'clock from reference point."
(d)	Announce target.	"Target, a cavalry patrol."
(e)	Announce range.	"Range, 1000."

1. All look to left front (or at 11 o'clock), and locate reference point.

2. A vertical clock face with its center on reference point is imagined. The men look along the line leading from its center through 10 o'clock, and find cavalry patrol at 1000 yards from firing point.

This system is best adapted to a rolling or hilly country. It may, however, be modified for a flat unbroken country as follows: Instead of imagining a vertical clock face, imagine a horizontal one with its center at the reference point and its 12 o'clock line in a general direction perpendicular to the firing line. Then proceed as described above.

Unit System. The units used are sight leafs, fingers, or mils in groups of 50. This system is suitable for indistinct targets, or sector designation.

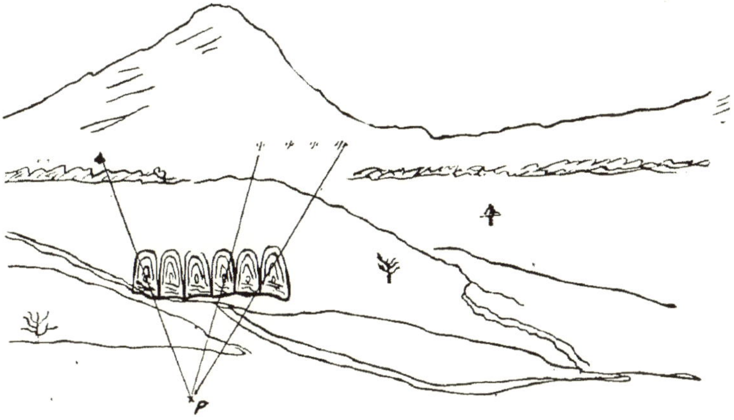

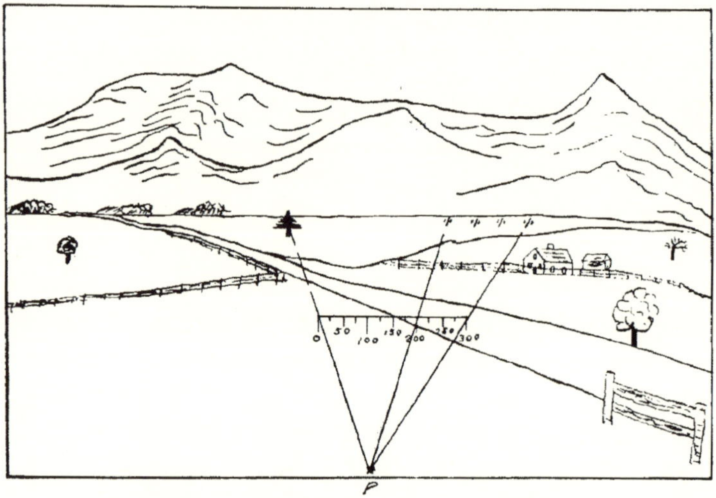

	METHOD	EXAMPLE
(a)	Indicate direction of reference point, as before.	"Reference point to our right front."
(b)	Describe reference point.	"Lone pine tree."
(c)	Announce angular distance, and, if necessary, direction from reference point to target.	"At 4 sight leafs" or, "At 4 fingers" or, "At 200 mils" to the right.
(d)	Announce target.	"Target, a machine gun company."
(e)	Announce range.	"Range, 800."

Procedure. 1. All look to right front and identify the reference point. 2. Then lay off successively to the right four widths of the sight leaf, or four fingers, or four units of 50 mils. This indicates the distance to the target.

3. Find a machine gun company at 800 yards from the firing point.

It is usual to indicate to the nearest point of the target. The distance the target extends beyond the initial point, as well as the division of the target into sectors may be given if desirable, as for example, "The target extends two fingers farther to the right. Each platoon will cover one finger of its own front."

When used in this way, the sight leaf or finger should be held 15 inches from the eye.

The unit system is probably the best method of target designation, but it is necessary that a definite unit be used. The most convenient unit is the mil, and it is absolutely necessary for those interested in machine guns to make themselves familiar with the mil system, since the field glass now issued to the service is equipped with a mil scale, and since the musketry rule is coming into general use. A little study of this system will be profitable, as it can be used for many things, such as target and sector designation, calculation of occupied fronts in terms of yards or mils, estimation of ranges, calculation of sight settings, windage corrections, and indirect fire data.

Discussion of the mil system. The mil is the angle whose tangent is 1/1000; that is, it is the angle subtended by 1 unit at a distance of 1000 units, for example, 1 yard at 1000 yards, or 1 foot at 1000 feet. In the conventional angular measure, its value is 3′ 26.2″ or about 1/18th of 1 degree. The mil never varies in value, but of course subtends greater widths at increasing distances, as ½ yard at 500 yards, 1 yard at 1000 yards, etc. This is illustrated in the sketch.

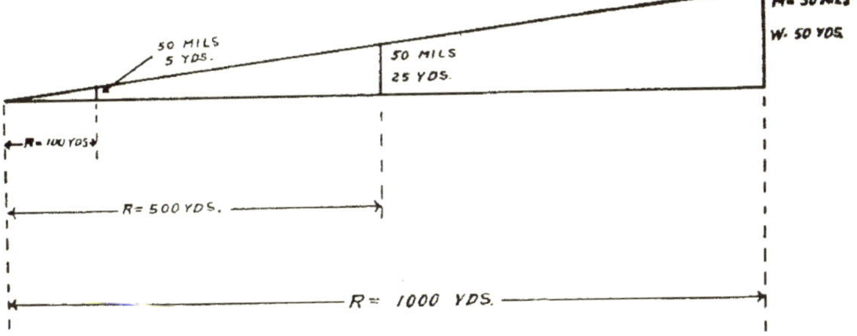

It should be no more confusing to be told that the ends of an enemy trench are 50 mils apart than it would to be told that they subtend an angle of 2° 32′ 11″, which is the same thing. Mils are measures of angles, just as degrees, etc., but are much simpler, as instead of being given in relation to circular measure, they are given in direct relation to range and width, so that calculations are facilitated.

One inch at a distance of 20 inches from the eye subtends 50 mils, (or 1/20th of the range). The average finger must be held 15 inches from the eye to subtend 50 mils.

From the definition of a mil and the principle of similar triangles, the following equations are evolved:

Let R=Range in Yards.

W=Width or height of target in yards.

M=Number of mils subtended by W.

$R \times M = 1000 \times W$, or $RM = 1000\ W$.

Transposing, $R = \frac{1000\ W}{M}$

$$M = \frac{1000\ W}{R}$$

$$W = \frac{RM}{1000}$$

By simply remembering that $RM = 1000\ W$, it is a short mental calculation to determine any one of the three terms, knowing the other two.

The mil rule. A conveniently simple and cheap form of mil rule may be constructed as follows: Fasten a 20 inch string to a lead pencil upon which notches have been cut 1 inch apart. Upon holding a knot in the end of the string to the eye, or in the teeth, the distance between each two notches represents an angle of 50 mils, (or 1 finger) and will subtend at all distances 1/20th of the range.

A still better rule may be fashioned out of white cardboard or celluloid, graduated in inches and tenths of inches. Such a rule held 20 inches from the eye subtends 50 mils for each inch.

By so graduating transparent celluloid, an unusually accurate rule is obtained, since the eye looks between the graduations instead of over them.

To calculate the graduations for any rule, simply substitute in the equations, using inches for R and W.

Example, let R be 20 inches and W be 1 inch. To find out how many mils this graduation represents at this distance, proceed as follows:

$RM = 1000\ W$. Transposing, $M = \frac{1000\ W}{R}$

Substituting, $M = \frac{1000 \times 1}{20}, = 50,$

or, 1 inch at 20 inches represents 50 mils.

The practical uses of a mil rule are as follows:

Estimation of range. A railway water tank is known or estimated to be 10 yards high. It is desired to know the range to the water tank, as there is a machine gun near it. Its height, read upon the mil stick is found to be 10 mils. What is the range?

$RM = 1000\ W$, or $R = \frac{1000\ W}{M}$

Substituting, R = 1000 yds.

Estimation of fronts. A troop of Cavalry in column of twos is crossing the front of a machine gun at a range of

1100 yards. From head to rear, the column measures 90 mils. How many troopers are in the column?

$$W = \frac{RM}{1000} = \frac{1100 \times 90}{1000} \text{ or 99 yards.}$$

Allowing one trooper per each two yards, the column will contain 50 men.

Sector Designation. A machine gun company of four guns, whose left flank is the only part of the company visible, is under cover at 1200 yards. Allowing a maximum interval of 50 yards between guns, what will be the distance to the right, in mils, to be covered?

The distance to be covered will be 150 yards.

$$M = \frac{1000\ W}{R} = \frac{1000 \times 150}{1200} = 125 \text{ mils.}$$

Determination of Ranges where the height or width of the target is unknown. First, read the number of mils, M_1, subtended by the target. Second, go forward or backward a certain known distance, D, and again read the number of mils, M_2, subtended by the target. Substitute the data thus obtained in the formula, Range = Distance of advance or retirement, times Second mil reading, divided by Difference between mil readings, or $R = \frac{DM_2}{\text{difference of mil readings.}}$

This formula is derived from the previous one by factoring and eliminating W.

Example. From a certain point a hostile trench is found to measure 200 mils. On advancing 200 yards, it is found to measure 250 mils. What is the range?

$$R = \frac{DM_2}{\text{Diff. in mils}} = \frac{200 \times 250}{50} = 1000 \text{ yds.}$$

This formula is equally correct whether the distance D be an advance or a retirement. An easy way to remember this range formula is to note that regardless of which way D is measured, the numerator of the fraction is the product of the last two operations performed, i. e., measuring or pacing D and making the second mil reading; while the denominator is always the difference between the two mil readings.

The mil rule as a range finder when the initial range is known: The most accurate use of the mil rule as a range finder is where the initial range is known, thus allowing the width of the target to be determined in yards. As the firing line advances, it is then necessary to make but one mil reading for each advance, and a simple calculation immediately gives the range.

Example: The flanks of an enemy trench are seen to subtend 120 mils when the firing line begins its deployment at a known range of 1200 yards. Hence the width of the enemy trench in yards is $W = \frac{RM}{1000}$, or $W = \frac{1200 \times 120}{1000} = 144$ yards.

As the firing line goes forward, it is found at the first halt that the mil reading on the same points on the flanks of the trench is 150 mils. What is the new range?

$$R = \frac{1000\ W}{M} = \frac{1000 \times 144}{150} = 960 \text{ yards.}$$

At the second halt the reading is 160 mils. What is the range?

$$R = \frac{1000 \times 144}{160} = 900 \text{ yards.}$$

At the third halt, if the reading is 180 mils, the range would be $R = \dfrac{1000 \times 144}{180}$, or 800 yards.

In this way, regardless of the number of halts, the range is read immediately by using 1000 × W as a constant, (as long as the target is the same), and dividing by the mil reading, an operation which could hardly be simpler.

A slide rule, constructed by laying off graphically the range in yards, the width in yards, and readings in mils, on a scale, will automatically give any one of the three factors, the other two being known; but the calculations are so easily made that with a little practice no such aid is necessary.

Range Maps.—The British use two types of range maps for machine guns, one being adapted for the offence and one for the defence.

Before an advance is made, the range finder gets ranges to all points of military importance, calculates his data, and constructs a map for the offensive. An example of such a map is shown in the sketch. The range finder selects suitable positions for machine guns on the line of the probable advance, and records the range to each position, on the right hand margin of the map. After taking and recording all ranges, he calculates the distance of each of these points from the objective, crossing out the ranges on the right hand side of the map and writing the ranges to the objective along the left hand side.

This map, upon being given to the company commander, permits him to read off quickly the correct range to the objective at each of the various halts during the advance, by simply glancing at the left hand margin.

For the defense, the map is made as shown in the sketch. The double line, P A, is the sighting line towards the probable front, which allows the map to be quickly oriented.

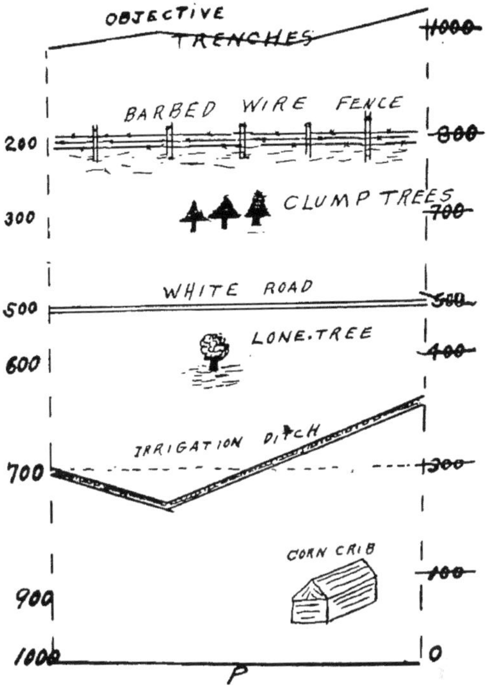

Range map for the offensive

Machine gun Signals. The use of a good simple system of signals is a necessity if the machine guns are to be handled efficiently in action or preparation for action.

The prescribed arm and whistle signals, as given in the Infantry Drill Regulations, Field Service Regulations, and Machine Gun Drill Regulations, should be consolidated on a single sheet, which each officer and noncommissioned officer of the machine gun company should paste in the back of his field notebook, ready for reference if at any time he should forget them. In addition, a few other signals will be found necessary, and they should not be improvised at the last moment, but should be carefully thought out beforehand, with

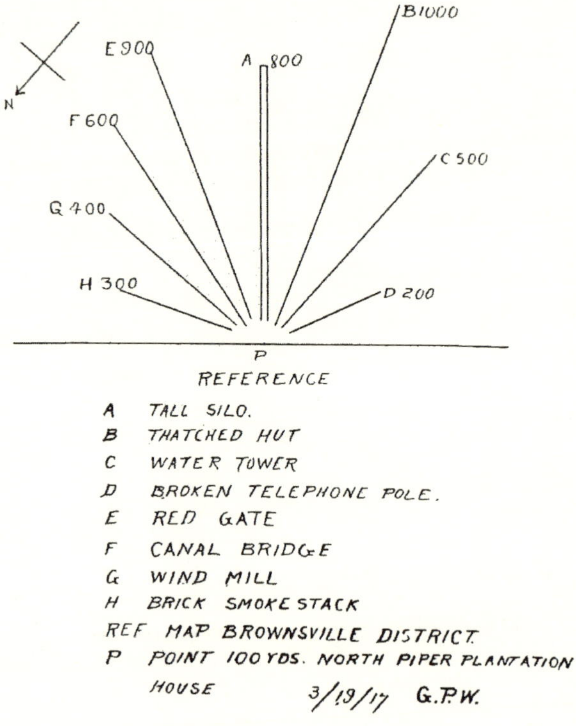

Range map for use on the defensive

the object of making every signal mean something. In devising signals, it is not sufficient to arbitrarily select a movement and call it the signal for a prescribed action; the motion used must, if possible, be so selected as to express the idea to be conveyed, otherwise the signals will be confusing.

On account of the liability to changes and amendments in the various manuals prescribing signals, it is not deemed expedient to list a set of signals here. It is suggested, however that signals be devised for the following movements:

SIGNAL	KIND OF SIGNAL
A. Combined sights.	Arm or Hand.
B. Searching Fire. (Hundreds of yds.)	Arm or Hand.
C. Windage.	Arm or Hand.
D. Bring Company forward.	Whistle.
E. By hand to the front.	Whistle.
F. Squad column.	Arm signal.
G. Platoon column.	Arm signal.
H. Extend.	Arm signal.
I. Close.	Arm signal.

Signals D and E are used in cases where the company commander has gone forward to reconnoitre, and desires either to bring up his company and animals, or to bring up the company and leave the animals in the rear. As the company commander may be separated from the remainder of his company by several hundred yards of underbrush or mesquite, it is ordinarily quite impossible for him to use visual signals or take time to send back a messenger.

A system of fire control signals similar to those in use in the field artillery should also be devised, for at present the use of indirect fire with machine guns is not infrequent.

Fire control signals should be devised for the following:

(a) Fire observed not less than 50 yards beyond the target.

(b) Fire observed not less than 50 yards short of the target.

(c) Range is correct.

(d) Fire observed to the right of target.

(e) Fire observed to the left of target.

(f) Deflection is correct.

(g) Unobserved.

(h) Fire observed, uncertain.

(i) Out of action.

These signals are usually letters or combinations of letters sent by visual signals. The letters used for each signal should be significant of the information to be conveyed, as for example, R R (Right Range) for "Range is Correct."

CHAPTER VI

MACHINE GUN MARKSMANSHIP

This chapter outlines methods which have been found to be of value in training the individual members of the organization in so handling the gun as to increase the effectiveness of the fire.

Two distinct kinds of training are necessary. These are classified as shown below, and will be treated in this order:

I. Exercises which do not involve firing; position and aiming drills.

(a) Sighting Exercises.

(b) Simulated fire at Miniature Range targets, involving exercises in the use of the elevating and traversing mechanism.

II. Firing Exercises with Ball Cartridges.

(a) Miniature Range Practice.

(b) Known Distance Practice. (300 and 500 yards.)

(c) Field or Combat firing.

POSITION AND AIMING DRILLS:

(a) *Sighting Exercises.*—The first step is to teach the gunner the correct position behind the gun and the correct method of holding the piece. After this has been learned, he should be taught the method of sighting and of setting the sight.

For this training, sighting triangles should be made, as prescribed in the Small Arms Firing Manual. This work can be facilitated by the use of a pantograph attached to the rear sight of the gun. The theory of the pantograph is fully described in the Encyclopaedia Britannica, and the application of this instrument to the machine gun is shown in the accompanying sketch. It is used to make a magnified record

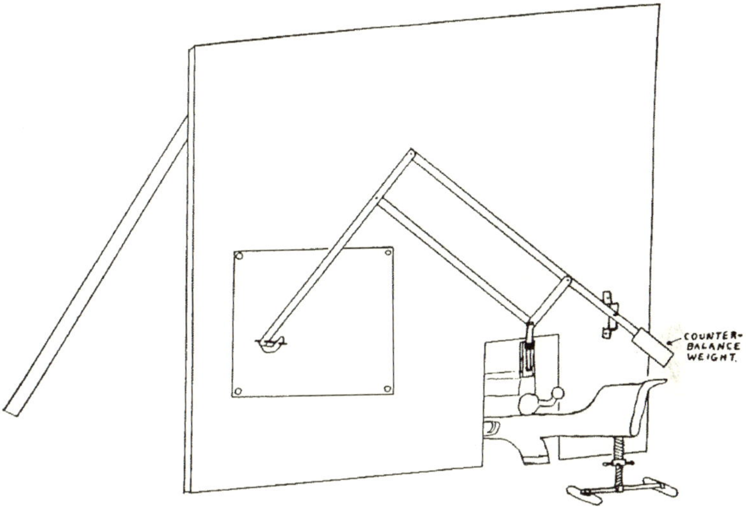

One method of applying pantograph to machine gun

of the travel of the point of aim in sighting exercises, and while it is not essential, its use will be found to be a great help.

Sighting triangles should be made with the machine gun resting on its own mount, if it has one. This drill should be conducted with the target at least 100 yards away, in order to teach the gunner to focus his eye on the target and not on the front and rear sights. At this distance a four inch bullseye should be used.

The use of a pantograph for this exercise will much simplify the work, as the triangles are made on a piece of paper close to the gun.

Exercises in the use of the Elevating Mechanism.—Place a target at about 28 yards from the firing point and locate pasters on it thus, numbering the pasters so that the numbers can be seen from the firing point:

.1 .2

.3

.4 .5

Then require the gunner to shift his aim from one point to another at the command of the instructor. Require him to obtain accuracy first, then speed.

(b) *Simulated fire at Miniature Range Targets.*—For Miniature Range work a distance of 28 yards is recommended, for the reason that at this distance a mil is equal to one inch. This appreciably simplifies the work of measuring shot groups and centers of impact, and makes the determination of the soldier's personal error in mils exceedingly convenient, since mils and inches on the target are the same.

For both simulated and actual firing the following targets are recommended:

Four targets, called No. 1, No. 2, No. 3, and No. 4, are used. These targets are all of the same size and shape, but differ in the arrangement of the aiming points. The targets are made in sections 3 feet high by 6 feet long. Four sections, or 24 feet, is the normal length of a target, but one or more sections can be used alone if it is desired. To construct the target, make four frames 3 by 6 feet, and cover them with target cloth. Then take the regular 300 yard target, cut it in half, and paste the paper thus obtained on the target cloth so that the plain side shows. Then add the necessary aiming points by pasting on single black pasters in the required location.

To facilitate measurements, divide the target by drawing vertical lines 1 foot apart. Then draw three horizontal lines, one six inches from the top, one in the middle, and one six inches from the bottom.

Target No. 1.—Each section of this target has three vertical lines of three pasters each. The pasters are placed on the intersection of the first, third, and fifth vertical lines with the three horizontal lines.

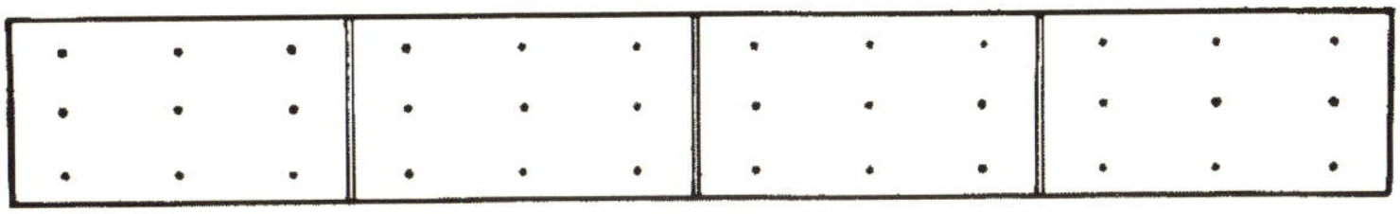

Target No. 1

This target is used for "fixed fire," that is, fire directed at a single aiming point. It is also used for shot group practice and for exercises in the use of the elevating mechanism.

Target No. 2.—Each section of the target has three horizontal rows of 24 aiming points per row. Each foot of each of the three horizontal lines has four aiming points spaced three inches apart, the first and last aiming points of each foot

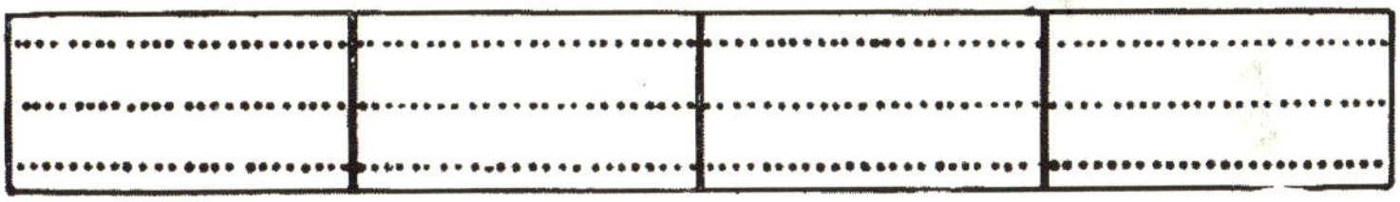

Target No. 2

being 1½ inches from the vertical lines. This arrangement of pasters is designed to represent infantry advancing in successive skirmish lines, or "waves." Allowing a density of one man per yard, the pasters represent a skirmish line at about 325 yards.

Target No. 3

Target No. 3.—This target is designed to represent troop movements in columns over hill sides. Each section has an inverted V shaped figure made by a line four inches wide composed of black pasters. The center of the figure rests on the intersection of the third vertical line with the top horizontal line. The lower ends of the figure rest on the intersection of the lower horizontal line with the first and fifth vertical lines.

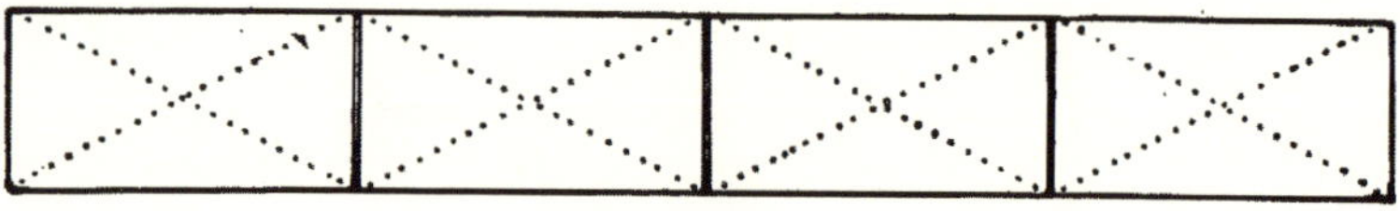

Target No. 4

Target No. 4.—This target is designed to represent a skirmish line in various positions. Each section consists of two diagonal lines of twenty aiming points per line, drawn so as to connect opposite corners of the section. The aiming points are located at four-inch intervals along the diagonals, the first and last pasters on each diagonal being three inches from the corners of the section.

Allowing a density of one man per yard, the pasters on this target represent a skirmish line at a range of about 250 yards measured perpendicular to their front.

Exercises in Searching Fire. Searching fire is fire distributed in the direction of depth.

Conclusive tests abroad and at home have demonstrated that it is the most effective fire that can be delivered from a machine gun.

Some means must be devised to control the amount of searching fire used. This can easily be done with all guns having a mount, and is particularly easy if the mil system is used.

With the Benet-Mercié gun, one turn of the outer elevating screw is equivalent to 16 mils, so that $\frac{1}{8}$ turn is equivalent to 2 mils. This simply means that for all ranges the line of the bore is elevated 2 mils for each 1/8 turn of the handwheel down. It follows that the strike of the bullet is also raised the same amount, which in this case would be 1 yard at 500 yards, 2 yards at 1,000 yards, etc.

Gunners should have some idea of how far a turn or a fraction of a turn of the handwheel will throw the bullet at any range.

A table giving this information may be constructed from a consideration of the value in mils of each turn of the handwheel, taken in connection with a table of angles of departure in mils.

Using target No. 1, exercises in lowering and raising the line of sight should be practiced. The command would be, "Searching fire (so many) mils." Or, "Searching fire (so many) turns of handwheel."

Example.—Suppose that the order "Searching Fire, 4 mils" is given. The gunner would bring his sight to bear on a paster and would then turn his elevating handwheel one quarter turn down (Benet), bringing the line of sight 2 mils above the paster. Then he would raise the handwheel one-quarter turn, which would bring the line of sight 2 mils below the paster. The gunner should be required to locate a spotter at the highest and lowest points of his aim. If the exercises have been correctly performed there will be an equal distance between the paster and the two spotters. This distance on the 28 yard target would be 2 inches, equivalent to 2 mils. Thus an accurate check of the amount of travel can be obtained.

This drill can also be held at 100 yards, in which case the distance between the paster and the spotters would be 7¼ inches.

Exercises in Distribution.—(Target No. 2.) This exercise consists in simulating aimed fire at each figure, going first from left to right, then from right to left.

Exercises in Searching with Distribution.—(Target No. 2.) This is an exercise that combines searching fire with distributed fire. The command is "Searching fire with distribution, (so many) mils."

The gunner aims at the first paster in the line and searches through the vertical distance indicated in the command. He then aims at the next paster in line and searches, and so on, his point of aim describing a line about as shown in the diagram.

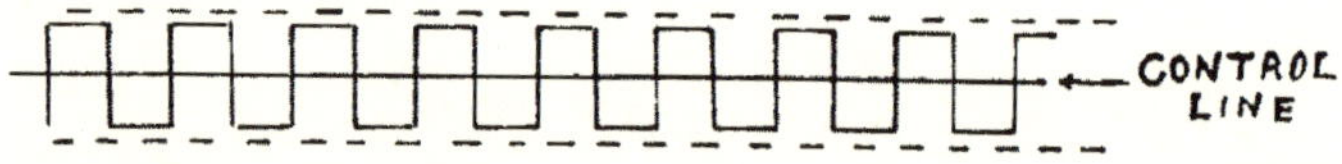

Exercises in searching with diagonal distribution.—(Target No. 1.) This is similar to the last exercise, except that the distribution is done on diagonal lines. For example, after searching the first point of aim, the fire is directed at the lower aiming point in the next row to the right, then at the upper aiming point in the next succeeding row, and so on, diagonally upward and downward until the edge of the target is reached. The process is then reversed. The line of sight would travel about as shown in the diagram.

The aiming points should be approached from below, and no attempt should be made to work the elevating mechanism simultaneously with the traversing movement.

II. FIRING EXERCISES WITH BALL CARTRIDGES:

(a) Miniature Range Practice.—For this practice targets 1, 2, 3, and 4, described above, are used on a 28 yard range.

The practice includes:

1. Fixed Fire.
2. Searching and Distribution.
3. Night Firing.

These classes of firing will now be described.

Fixed Fire.—The object of this practice is to teach the gunner to make good shot groups and to determine his personal error in aiming. The practice is carried out as follows:

The gunner is allowed to fire bursts of five shots at designated pasters on No. 1 target, and the resulting groups of shots are gauged with an ellipse of wire whose horizontal diameter is 3 inches and whose vertical diameter is 5 inches. This practice should be continued until close groups are obtained, after which the gunner should be allowed to fire longer bursts until he can fire the entire strip and still secure a close grouping. Stray shots are not counted. The remainder of the group should be close enough to allow it to be covered by the ellipse.

To determine the soldier's individual error, or personal equation in aiming, determine the centers of impact of his groups and compare them with the theoretical position for the sight setting which is being used. The theoretical position of the center of impact is given in table 1. The difference in inches between the theoretical position and the actual position of the center of impact is the soldier's personal equation in mils. This should be obtained for several sight settings, say 300, 700, and 1000 yards, and the results recorded.

TABLE I

Theoretical height of center of impact above the aiming point for various sight settings at a reduced range of 28 yards.

Sight Setting	Height of Impact, inches or mils
0	—.17
100	+.53
200	1.34
300	2.24
400	3.31
500	4.47
600	5.87
700	7.47
800	9.33
900	11.48
1000	13.93

This was derived from the table of angles of departure in mils, given in Table II, considered in connection with the fact that the drop of the bullet at 28 yards is .17 inch.

This table does not take into consideration the fact that the line of sight is above the bore of the gun. In making measurements, this distance must be considered. For the Benet gun the front and rear sights are 1″.72 above the bore at a sight setting of 100 yards. Therefore, at this sight setting the bullet would hit 1.72 below the point determined in Table I, and for 0 sight setting, it would hit .7 inch lower than this, or 2.4 below the line of sight. Therefore, all measurements must be made from a point 2.4 inches below the line of sight for the Benet gun. For the Maxim the zero is 4.83 inches below, and for the Vickers it is 4.50 inches below the aiming point. For the Lewis, using British ammunition, it is 3.4 inches below the aiming point.

The next step is to determine the correction that the soldier must apply to overcome his personal equation. This correction is conveniently determined in terms of the elevating screw head on the rear sight. One turn of this screw is equivalent to one mil. The correction to be applied can be determined by measuring the distance between the actual center of impact and its theoretical position. This distance in inches on the 28 yard target is the personal equation in mils, and can be corrected by turning the elevating screwhead the same number of turns that there are mils in the error.

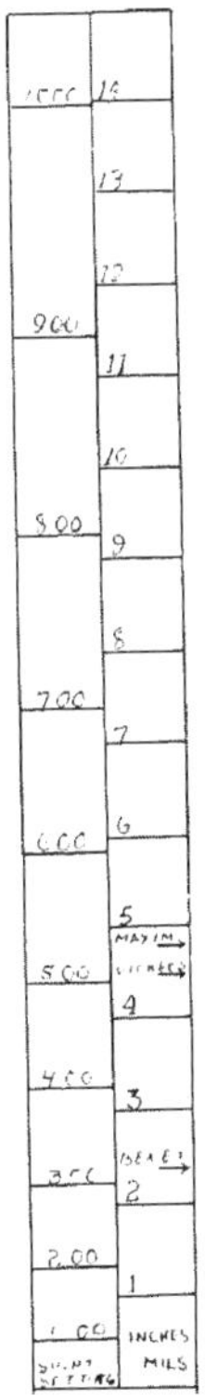

Scale of ordinates for Benet, Maxim, and Vickers (Quarter size)

Where a large number of such determinations are to be made, it is an advantage to construct a scale like the one in the figure, which is a graphic representation of the table of ordinates.

The rule is constructed of paper, wood, or cardboard, by graduating one side in inches, which also represent mils on the 28 yard target, and graduating the other side with the theoretical heights of impact, as given in Table I.

The zero for each type of gun is then marked on the rule.

To determine the soldier's correction with this rule, proceed as follows:

Place the zero of the rule on the aiming mark. The center of impact should then be at the height of the graduation corresponding to the sight setting used. The amount in inches that the center of impact lies above or below this point is the soldier's personal equation in mils, and this error will be corrected if, after the soldier sets his sight, he turns the elevating screwhead of the rear sight a corresponding number of turns down, or up, as the case may require.

Example.—Suppose that the gunner, with a sight setting of 1000 yards, made a shot group whose center of impact is above the aiming point.

Place the zero of the rule on the aiming mark. Suppose that the center of impact comes on the 11 inch mark on the right side of the rule. It is at once seen that the shot group is 3 inches too low. Therefore, the gunner should raise his sight 3 mils, or 3 complete turns of the rear sight elevating screw head.

The personal error of the soldier should be determined for at least three ranges, and the results averaged. This correction should then be applied to every sight setting used for target practice or action, by turning the screw head the required number of turns after the true range has been set on the sight.

The reason for making the correction in mils instead of yards, is that the error is a constant angle at all ranges, while the graduations representing yards on the sight leaf vary with the range, the graduations being more widely spaced as the range increases.

After practicing sight setting and correction of personal error, the gunner fires a few rounds to verify the correction, using target No. 1.

Practice in Searching and Distribution.—After the completion of the exercises in Fixed Fire, practice in Searching Fire, Distributed Fire, and combinations of the two are carried on with ball cartridges, following the system used in the position and aiming drill exercises. Some system of scoring for the practice should be devised. Simplicity should be the keynote of any such system, for the confidence of the men in the results will depend largely on whether or not they can understand how the results are arrived at. Any system that is adopted should be based upon the three important elements of effective machine gun fire. These three elements are, volume of fire, distribution of fire, and accuracy of fire.

Volume of fire is the rate at which fire is maintained; that is, the number of aimed shots fired per minute. It is represented by the relation between the total number of shots fired per minute and the number of aimed shots it is possible to fire in a minute from the type of gun used.

Distribution of fire is represented by the relation existing between the total number of targets hit and the total number of targets exposed.

Accuracy of fire is represented by the relation between the total number of hits made and the total number of shots fired.

In the preceding exercises, firing should be with single shots until proficiency is obtained, after which automatic fire may be employed.

The following two examples show the type of exercises that may be used:

1. The gun squad is given 300 rounds of ammunition, and fire is opened on target No. 4. "Cease Firing" is not given until the ammunition is exhausted.

2. The gun squad is given "Commence Firing." After one minute, "Cease Firing" is given.

Further practice should also be held with targets arranged to represent the following formations:

Squad as skirmishers, platoon as skirmishers, squad column, platoon column, platoon advancing in thin lines, squad and platoon rushes, etc.

(3) *Night Firing.*—The object of night firing on the miniature range is to give practice in setting up the gun, loading, correcting stoppages, etc., in the dark.

(b) *Known Distance Practice.*—After the completion of the 28 yard course, some firing should be done at 300 and

500 yard ranges. This will give the gunner some idea of the actual ability of his gun to make hits at these ranges. It also teaches him to focus his eye on the target and not on the sights, as on these longer ranges the eye cannot be focused on the front sight, the rear sight, and the target at once as it can on the miniature range.

The gunner will have more confidence in his weapon after the results obtained on the miniature range are substantiated by his ability to make hits at the longer ranges. A fair grouping of shots at 500 yards is far more satisfactory to the gunner than a perfect group at 28 yards.

At the 500 yard range the wind also becomes a factor, and practice at this range gives valuable practice in making windage corrections.

Practice at 300 yards.—(Target A, S. A. F. M.) The gunner fires thirty single shots, taking the necessary time to enable him to correct his sights and get in the bullseye if possible.

Then with automatic fire he will fire 5 feed strips (150 shots) in one minute. (Benet)

Score will be the total number of points made by the firer.

Practice at 500 yards.—(Target B, S. A. F. M.) The gunner fires thirty single shots, using the leaf sights; then with the telescopic sight properly adjusted in accordance with the instructions in the handbook, he will fire thirty single shots, using the telescopic sight. If necessary, the adjustment for range will be corrected so as to get hits in the bullseye with the range dial set for 500 yards.

After the adjustment of the telescopic sights has been verified by firing, each gunner will fire automatically five feed strips (150 shots) using the leaf sight. Time allowed, 1 minute. He will then fire the same exercise, using the telescopic sight. The score for each exercise will be the total number of hits made. After each exercise the target is marked so that the gunner can see just where the hits were made.

(c) *Field or Combat Firing.*—After the completion of known distance practice, machine gun training should be under simulated service conditions. For the field firing a terrain should be selected where the firing will not endanger life or property, and the targets should be located in groups of varying sizes and shapes representing a variety of military formations. Care should be taken in the location of these targets, and they should be so placed as to secure fair concealment and cover. Only two types of target need be used; the prone olive drab silhouette and the kneeling olive drab silhouette, targets E and F, S. A. F. M. Short tent pegs will easily hold the prone, and long tent pegs the kneeling figures.

Before the firing of the exercises, the targets are placed in position under the supervision of an officer. The target detail is then withdrawn, leaving the range ready for firing.

The various exercises should be fired first by squads, then by platoons, and finally by the entire company.

Sometimes reconnaissance by machine gun scouts or squad leaders for the purpose of locating the targets will be allowed. At other times the general location of the targets for a special exercise is indicated to the commander of the unit to fire. At still other times the exercise is in the nature of a surprise and the action must be initiated instantly.

Ranges and data are always unknown. All movements are made by command, and service conditions must be maintained.

Ranges should vary from 2000 yards down to 300 yards. Indirect fire should be practiced considerably. If the terrain is flat, the targets can be screened from the guns for the purpose of practicing indirect fire. Some exercises should be staged during the day and fired at night as night firing exercises.

A machine gun company exercise should be varied from going into action from a marching formation to advances by hand, with an assumed firing line over various types of terrain and against an imaginary enemy whose forces are considered to consist of infantry alone, or of infantry and artillery.

There are an indefinite number of situations which can be worked out by an energetic company commander, and such exercises are of inestimable value for practice in estimating distances, target designation, extended order drills, fire control, machine gun marksmanship, etc. A company that cannot conduct itself well under these conditions is not well trained. The amount of ammunition used need not be large if care is used in devising the problems.

Targets are always marked after the completion of each exercise, and the scoring is recorded.

CHAPTER VII

INDIRECT FIRE

Under this heading not only indirect fire, but also night fire and overhead fire will be discussed.

Indirect fire with machine guns is not only practical, but is often a necessity. It is used to cover restricted areas over which an enemy may be expected to advance, such as bridge heads, fords, ferries, defiles, cross roads, ditches, etc.

It is also used to search distant dead spaces, hollows, woods, etc., particularly to locate an enemy.

The chief advantage of indirect fire is that it is screened from direct hostile fire. An enemy, upon being subjected to indirect fire, becomes more dismayed than if direct fire were used, for he cannot see the hidden machine guns that are decimating his ranks, and is consequently unable to return an effective fire.

Upon the machine gun crew, indirect fire produces a feeling of security and confidence, for they are hitting the enemy without being seen, and consequently, without being hit.

On account of the calmness of the gun crew, and because the fire is more mechanical, indirect fire is often more effective shot for shot than direct fire, and will often produce more hits.

The chief disadvantage of indirect fire is that it is slow, and requires skill and preparation to enable it to be effective. Its chief use is, therefore, on the defensive, where it is invaluable.

The following discussion gives two examples of the correct tactical use of indirect fire.

Example No. 1.—A machine gun company of four guns in the fork of a road at X, as shown in the sketch, can, by the

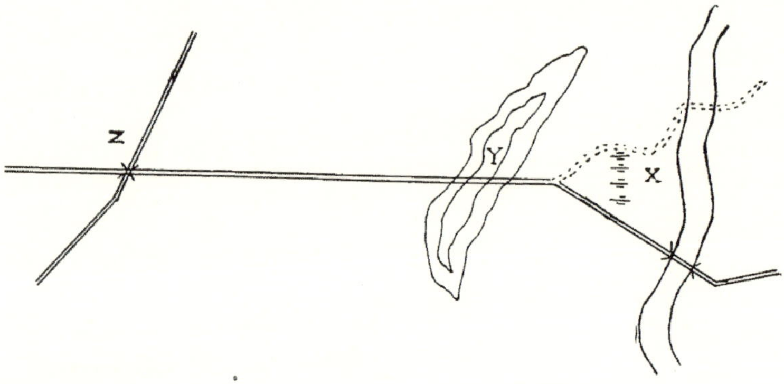

use of indirect fire over the crest of a slight rise in the road at Y, completely cover the bridge at Z, 800 yards away.

The advantages of the position at X, with indirect fire, over the position at Y, with direct fire, are:

1. Concealment.

2. Cover. (Since the enemy would not employ indirect fire in reply, from lack of time and preparation.)

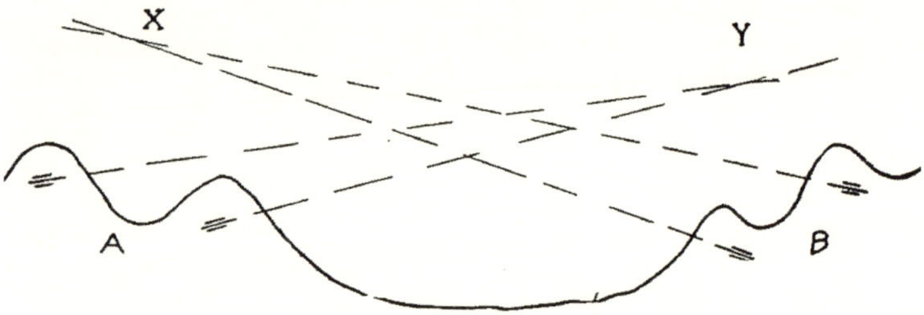

Example No. 2.—Referring to the sketch, the position of the advancing line of enemy is X Y.

Position of our machine guns, behind folds of hills, **A** and **B**.

The advantages of such a position are:

1. Flanking fire, as A sweeps Y and B sweeps **X.**

2. Concealment.

3. Cover.

4. It is impossible to locate the guns by sound, as the noise of the cross fire seems to be in front of the position.

A resourceful machine gun commander will know a number of methods for utilizing indirect fire, and will carry on his person constantly when in the field the necessary data and instruments for indirect firing. The apparatus need not be elaborate or complicated, as all that are required are, a thorough knowledge of the methods of indirect fire, a note-book containing tables and other data, a twenty-five cent pocket level, and a mil rule, either in the field glasses or on a scale.

The principles of indirect fire are based on the fact that the bullet will always strike the target if the angle of elevation of the bore is correct and the proper deflection is used. Hence, with the bore of the gun properly directed, the sights of the gun can be altered, without moving the gun, so that any desired aiming point within the capacity of the sight leaf and windage scales may be used.

The following table outlines the most useful methods of indirect fire. Some of these methods are comparatively simple, while others require a considerable amount of preparation and skill. That method should be used which will give good results with the least amount of preparation, and long methods should not be used unless circumstances make it necessary.

There are but three general methods of indirect fire, though there are a variety of ways of using each method.

Methods of Indirect Fire:

I. Distant Aiming Points.

A. Graticules.

1. Inverted sight leaf scale.

(a) New EE type field glass, fitted with Eames scale.

(b) Inverted sight leaf, held at sight base length from the eye.

(c) Musketry Rule.

B. Mil Scale.

1. Elevating screwhead on rear sight.
2. Elevating Mechanism Handwheel.
3. Mil table of angles of departure.

II. Pocket Spirit Level.

A. Gun and Target on approximately the same level.

III. Quadrant Elevations.

A. Direction obtained by prismatic compass and elevation by clinometer.

B. Direction obtained by aiming posts and elevation by vertical angles.

1. Gun aligned by observation.

(a) One post.

(b) Number of posts.

Case 1. Distant Auxiliary Aiming Point:

A. By the use of a graticule (card, showing graduations of rear sight leaf, attached to a string), inverted sight leaf, Signal Corps field glass fitted with Eames (inverted sight leaf) scale, or with the musketry rule. Procedure:

1. Obtain true range to the target.

2. Secure a position near the gun (not over six feet above it), from which the target may be seen. Sight at the target with the graticule or scale, placing the line of scale which indicates the true range so that it cuts the target. See sketch.

3. See which graduation on the scale crosses a suitable aiming point above the target and visible from the gun.

4. Order the gunner to set his sights to the range corresponding to this graduation. Upon opening fire on the

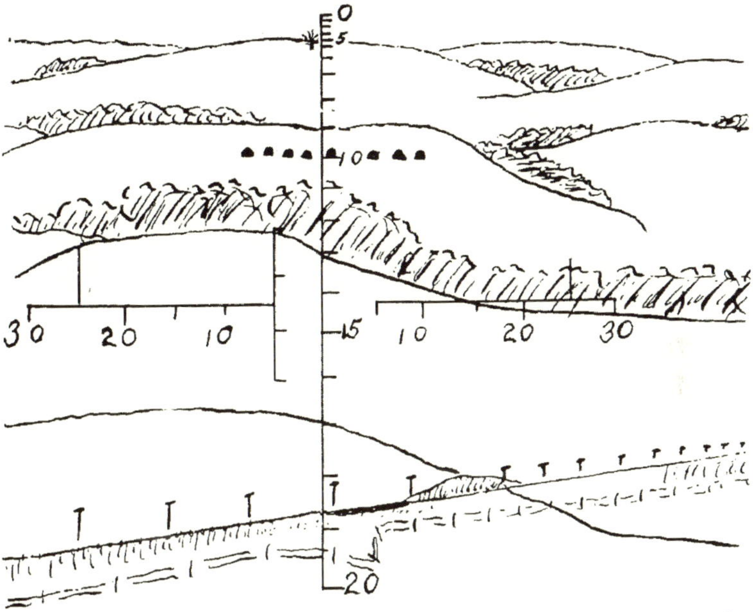

aiming point with this sight setting, the shots will strike the target.

Example: Range, 1000 yards. Upon placing the 1000 yard graduation on the target, the top of a distant tree on the skyline above the target is cut by the 400 yard line on the scale. Then, upon opening fire on the top of the distant tree, with the sight set at 400 yards, the shots will strike the target. This is shown in the sketch.

The aiming point has been considered to be vertically above the target when viewed from the gun position. If it is to one side, windage must be taken. The proper amount of windage is determined by measuring the deflection in mils and taking as many points of windage in the opposite direction. For all practical purposes 1 mil is taken to be equal to 1 point of windage, though the exact value is 1 and 1/6 mils for each point of windage.

If no mil scale is available, the windage correction may be determined as follows: Take aim, with windage zero, at a point directly above the target; then, without moving the gun, turn the windage screw until the sight is directed at the aiming point. This automatically corrects for the lateral distance of the aiming point.

An explanation of why the gun will hit the target at 1,000 yards when the aim is taken at a point above the target with a different sight setting is as follows:

Suppose the gun to be sighted at the true target with the correct sight setting of 1000 yards. It is obvious that the bore is then pointed correctly to send a bullet through the target. Now if the bore is left pointed this way and the sight setting is changed to a lower reading, the line of sight will be raised, since the gun is not being moved. When the line of sight has been raised enough to enable the distant tree on the skyline, in the example, to be seen, the rear sight will be at 400 yards. Now if fire is opened, the target will be struck, as the gun has not been moved. By inverting the sight leaf scale and reading direct, the same process can be carried out, as described in the example given above.

NOTE.—The following directions are for those unfamiliar with the use of the sight leaf for this purpose:

Obtain a rear sight leaf from the spare parts of the Springfield Rifle, and tie a good quality of fish cord, 22⅛ inches long, to it. Then by holding the knot on the end of the string in the teeth, the sight leaf can be held the proper distance from the eye to enable it to be used in the inverted position as a graticule. See sketch.

B. Mil Scale System:

Procedure.

1. Obtain the range to the target.

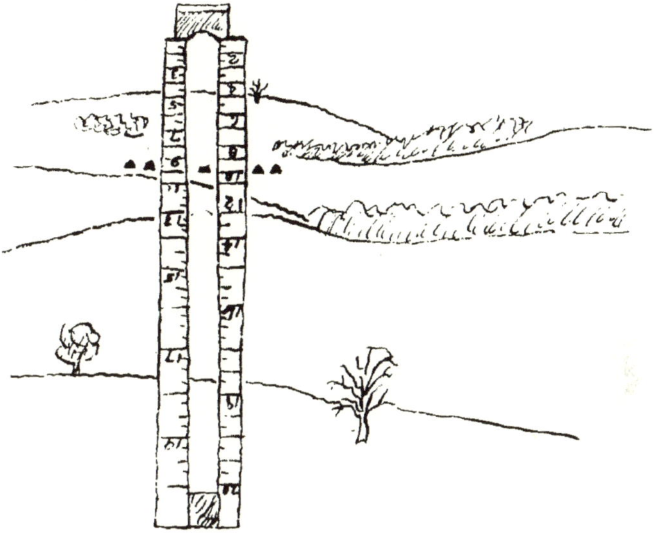

2. Take a position from which the target can easily be seen. Select a convenient aiming point above the target, such as the fork of a tree, and take the mil reading of the vertical distance between the target and the aiming point. Also, if the aiming point is not directly above the target, take the windage reading as described in case 1.

After reading the number of mils between the target and the aiming point, to get the proper setting, proceed with any of the following three methods:

(a) By the use of the rear sight elevating screw head.

3. Set the sight to the correct range to the target.

4. Then, by the use of the elevating screw head, move the rear sight down, making as many complete turns of the screw head as there are mils in the verticle distance between the target and the aiming point.

5. Open fire on the aiming point with the sight in this position, after making the proper windage correction as described above.

Example.—Range, 1000 yards. Vertical interval between the target and the aiming point, read in the field glasses, 10.5 mils. The aiming point is 5 mils to the right of the target.

Set the sight at 1000 yards, then turn the elevating screw head down ten and a half turns, which will bring the rear sight to 400 yards. Set off 5 points left windage, and open fire at the aiming point. The target will be struck.

This method is based on the fact that each turn of the rear sight elevating screw head is equal to one mil. Each notch is equal to ⅛ mil.

(b) By the use of the elevating handwheel.

3. Set the sights to the correct range to the target.

4. Aim at the aiming point.

5. Then with the elevating mechanism, depress the gun as many mils as there are in the vertical interval between the target and the aiming point. (With the Benet, each turn of the elevating handwheel is equivalent to 16 mils, or ⅛ turn is equivalent to 2 mils.) This causes the bore of the gun to be properly directed to hit the target. However, as the target cannot be seen, it is necessary to align the sights on the aiming point, as follows:

6. Reset the sight, without moving the gun, until the line of sight will strike the aiming point.

Example.—Range, 1000 yards. Mil reading between the target and the aiming point, 10.5 mils.

Aim the gun at the aiming point with 1000 yard sight setting. Then turn the handwheel up slightly more than ⅝ of a turn, which depresses the bore about 10.5 mils, as each ⅛ turn is equivalent to 2 mils. Then, without moving the gun, move the rear sight leaf down until the aiming point is seen when looking over the sights. If fire is now opened, the target will be struck.

TABLE II

ANGLES OF ELEVATION					
HDS. YDS.	DEGREES	MILS	HDS YDS.	DEGREES	MILS
1	0° 2.4′	0.7	16	2° 2.4′	35.6
2	0 5.2	1.5	17	2 18.7	40.2
3	0 8.3	2.4	18	2 36.1	45.3
4	0 11.8	3.4	19	2 54.9	51.0
5	0 15.9	4.6	20	3 14.7	56.5
6	0 20.7	6.0	21	3 35.3	62.3
7	0 26.1	7.6	22	3 57.7	69.0
8	0 32.5	9.4	23	4 21.7	76.1
9	0 39.8	11.7	24	4 47.1	83.4
10	0 48.3	14.0	25	5 14.1	91.2
11	0 57.8	16.8	26	5 42.8	99.3
12	1 08.5	20.0	27	6 13.4	107.9
13	1 20.3	23.3	28	6 45.8	116.8
14	1 33.2	27.0	29	7 20.4	126.4
15	1 47.3	31.1	30	7 57.2	136.4

1 MIL = 3′ 26″ = 3.437′, 1° = 17.7 MILS

(c) By the use of a table showing the angles of departure in mils. Such a table is given herewith. The angle of departure, whether given in mils or in degrees and minutes, is the angle to which the bore of the gun must be elevated in order for the bullet to reach a given range, or distance on the horizontal, when the gun is fired. The table was derived from the angles of departure given in the Ordnance Department handbook for the Springfield Rifle, considering 1 mil as 3.437 minutes.

This table can easily be carried in a notebook, and has other uses besides those indicated in this method of indirect fire.

Procedure when the mil table is used:

3. By an inspection of the mil table, find the angle of departure in mils for the true range to the target.

4. Subtract from this angle the number of mils in the vertical interval between the target and the aiming point. The remainder will show the correct angle of departure for the sight (not the gun) to be set at to enable the aiming point in question to be used.

5. By an inspection of the table, find out what sight setting this angle corresponds to, and set the sights to this range. Then set the necessary deflection on the windage screw as previously explained.

Example.—Range 1000 yards. Vertical interval between the target and the aiming point, 10.5 mils. The table gives the angle of departure for 1000 yards as 14 mils. Subtracting 10.5 mils gives a remainder of 3.5 mils, which is found by an inspection of the table to correspond to a sight setting of 400 yards. Set the sights at 400 yards and aim at the aiming point. On opening fire, the target will be struck.

Case II. Pocket Spirit Level Method:

A. When the gun and target are on approximately the same level.

1. Obtain the true range to the target.

2. Set the sights at zero and level the gun with the pocket spirit level.

3. With the gun level, send out a man with an aiming post or any other object which can be used as a sighting mark, and have the mark approximately aligned between the gun and the target by an officer standing in rear of the gun in a position where the target may be seen. If the mark is not exactly in line with the target, set off enough windage to correct the error, as described before. Then have the mark

placed at the proper height to be in the line of sight with the gun level and the sights set at zero.

4. Set the sights of the gun to the true range of the target, without disturbing the level position of the gun.

5. Leaving the sights at this setting, change the elevation of the gun until the sights again bear on the aiming point.

6. Open fire, and the target will be struck.

Case III. Quadrant Elevation:

A. Direction obtained by prismatic compass and elevation by clinometer, and laid off graphically by means of stakes.

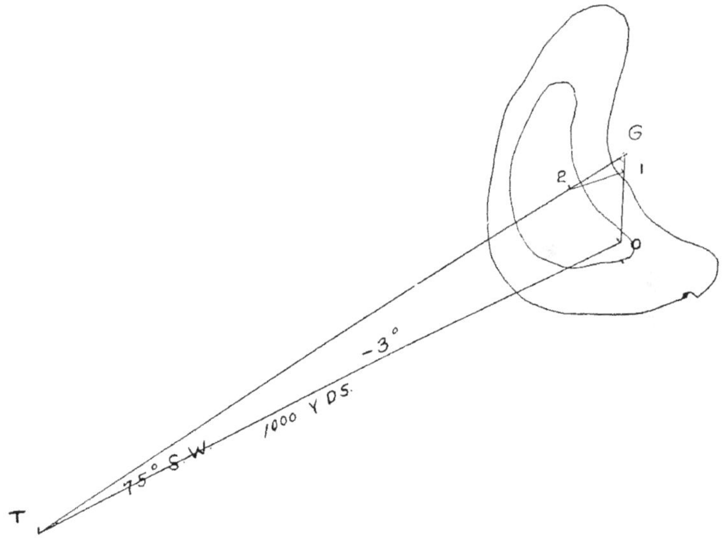

To understand this method, it will be necessary to refer to the accompanying sketches in connection with the following procedure. The machine gun is at G, target at T, and the observer, equipped with range finder, compass, and clinometer, is on the crest of the hill concealing G from T.

1. The observer takes the range to T, reads the compass direction of T, and with the clinometer, reads the angle of slope to T. He then signals this information back to the gun.

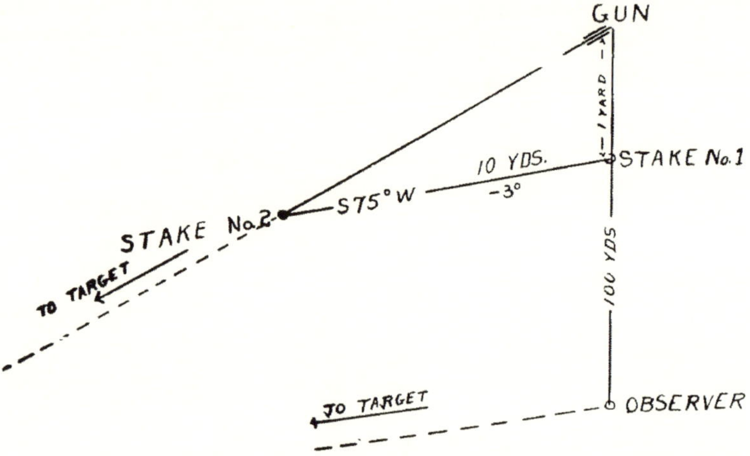

2. From the gun, the distance to the observer is obtained. In the figure, assume the distance from the gun to the observer to be 100 yards, the range from 0 to the target to be 1000 yards, the compass reading S. 75° W., and the clinometer reading to be 3°.

3. Drive a stake in line with the gun and the observer, 1 yard distant from the gun for each 100 yards the observer is distant. Call this stake No. 1.

4. Place the compass on Stake No. 1, and locate another stake S. 75° W., and as many yards distant from the compass stake as there are hundreds of yards in the range, from observation point to the target.

5. On the first stake, make a mark at the height that a line from the gun to the observer passes the stake. This

imaginary line indicates the slope of the plane containing the gun and the observer.

6. Holding the clinometer at this mark on Stake No. 1, take a reading of —3° on Stake No. 2, and make a mark. This is the aiming point.

7. Measure the distance from the gun to Stake No. 2, in yards, and multiply the result by 100. This gives the true range from the gun to the target.

8. Set the sight to this range, and aim at the mark on Stake No. 2.

9. To find other aiming points, shift the sights without moving the gun or disturbing the position of the bore. The gun is now ready for firing.

B: The Use of Aiming Posts.

This method is the most practicable one for the case where a hill intervenes between the gun and the target. To fire over an obstruction at an invisible target which is not on a level with the gun, it is necessary to obtain the direction and elevation of the target. The direction is obtained by the old familiar surveying method of "Lining in," or aligning stakes on the target, the near stake being placed close to the gun, to be used as an aiming point.

Since the gun and the target are not on the same level, a certain amount must be added to or subtracted from the angle of departure corresponding to the range. This difference can be obtained by the following method, which is based on the principle that the tangents of small angles are proportional to the angles.

In the figure, G is the gun, O is the obstruction on which an observer is located, and T is the target. Required, the angle of the target above or below the plane of the gun.

This is obtained by the following formula: Multiply the distance from the gun to the obstruction by the angle at O

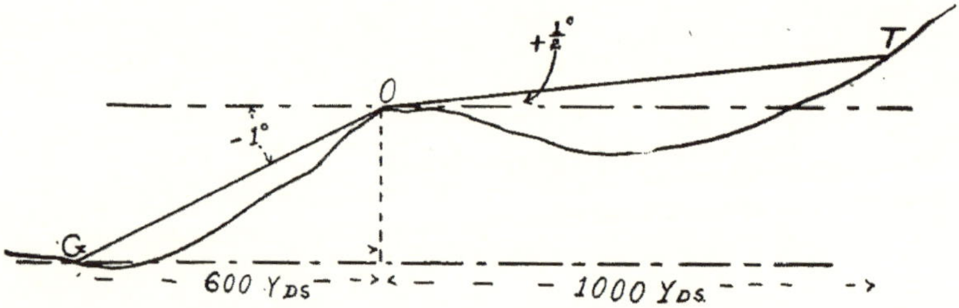

between the gun and the horizontal. Multiply the distance from O to the target by the angle at O between the target and the horizontal. Subtract the first result from the second, having due regard for algebraic signs, and divide the result by the total range. This will give the angle of the target above or below the gun, and must be added to or subtracted from the angle of departure for the true range. If the result is a plus quantity, the target is above the gun and the result must be added; but if it is a minus quantity, the target is below, and the result must be subtracted.

Example.—In the figure, let the distance from the gun to the obstruction O be 600 yards, the angle from O to the gun be minus 1 degree, the angle from O to the target be plus ½ degree, and the distance from O to the target be 1000 yards.

The total range will be 1600 yards. The angle of departure for this range is found from Table 2 to be 2° 2′.

This must be decreased by the angle of the target above or below the plane of the gun, which is determined as follows:

Substituting in the formula given above, we have,

$$\frac{1000 \times \frac{1}{2} - 600 \times (-1)}{1600} = +\ 11/16 \text{ degrees, or } 41'$$

Add this to the angle of departure found above.

2° 2′ + 41′ = 2° 43′, which the table shows to correspond to a sight setting of 1835 yards.

All that is necessary to hit the target is to line the gun with the target and fire it with an elevation of 2° 43′. The method of doing this will be described below.

Complete procedure for indirect firing by this method:

1. Choose the probable location of the gun.

2. Send the observer forward to locate a stake on the hill, approximately between the gun and the target. Then have a second stake placed in line with the first stake and the target, but nearer the gun. Then, if necessary, continue lining in stakes until one is conveniently located about 30 yards from the gun. If the gun is aligned with the stakes, it will be pointing at the target.

3. Have the observer at O read the angle from his position to the gun, prefixing the proper sign, plus or minus, according as the gun is above or below his position. In the example given above, the angle was minus 1 degree.

4. Have the observer read the angle from his position to the target, in this case, plus ½ degree.

5. Multiply each angle by its corresponding distance and divide the algebraic difference by the total range, as explained above. In the example, the result is plus 41 minutes.

6. Look up in Table II the angle of departure corresponding to the true range, and add (or subtract) the result obtained in (5). In the example given, 2° 2′ + 41′ = 2° 43′, the new angle of departure.

7. Look up in Table II the range corresponding to this new angle of departure. In this case, it is 1835 yards.

8. Level the gun with the pocket level, and with zero sight setting, place an aiming point so that it is in the line of sight, leaving the gun aligned in the direction of the target by means of the stakes, as described in (2).

9. Set the sights at the range obtained in (7), aim on the aiming point, and fire.

It does not matter whether the angles are read in mils, degrees, or minutes, as long as the same units are used throughout the calculation.

The foregoing method is particularly convenient when the Benet, or any gun with a similar rear sight, is used, for the use of tables of elevation can then be rendered unnecessary as follows:

First, set the total range on the rear sight, then take all angle readings and make all calculations in mils. The angle to be added to or subtracted from the rear sight reading will thus be obtained in mils, and can be added or substracted simply by turning the rear sight elevating screw-head up or down a corresponding number of turns, since 1 turn of this screw equals 1 mil.

The following methods of reading angles will be found of value when no clinometer is available. The necessary instruments are, a pocket level, Table II, and the gun.

To read an angle of depression:

(a) Level the gun with the pocket level.

(b) Without moving the gun, alter the sights until the line of sight is directed at the mark to which the angle is to be measured.

(c) Read the sight setting, and look up the corresponding angular value in Table II.

Example.—The sights read 100 yards when the line of sight is directed at the object. The angle is therefore 48.3′.

To read an angle of elevation:

(a) Level the gun with the pocket level.

(b) With zero sight setting, locate a point on the line of sight.

(c) Leaving the sight at zero, elevate the gun so that the line of sight is on the object whose angle is being measured.

(d) Then, without moving the gun, alter the sights until the line of sight cuts the level point, located in (b).

(e) Read the sight setting and look up in Table II the corresponding angle.

The use of the EE type field glass for reading angles.—Since this glass is fitted with a mil scale, the angle between any two objects within range of the scale may be read in mils, and converted into degrees or minutes, if desired. By fastening a small pocket level to the top of the glass with rubber bands, and using a small mirror held in one hand to reflect the bubble, the glasses may be kept level, and the angle of an object above or below the horizontal may be thus read. It is necessary in this case to first calibrate the glass to find out which point of the scale corresponds to the horizontal point when the glasses are leveled.

General Precaution for Indirect Firing.—When indirect fire is employed with the gun behind an obstruction, such as a wall, hill, etc., be sure that the height of the trajectory is sufficient to enable the bullets to clear the obstruction with the sight setting it is proposed to use. This may be determined as follows:

1. Aim at the aiming mark with the proper sight setting to cause the shots to strike the invisible target.
2. *Without moving the gun,* set on the rear sight the distance to the obstruction.
3. Look along the sights. If the line of sights clears the obstruction, the bullets will clear. If not, the gun must be moved farther back.

CORRECTION OF FIRE:

A ready method of correcting fire when the target is visible to the gunner, is for the observer (squad leader) to give the correction as a command, in terms of the apparent dimensions of the target.

For example, suppose that the target is a distant motor truck full of troops, and the observer has noted that the burst of fire was low, vertically, an amount equal to the apparent height of the truck, and was apparently twice the width of the truck to the right.

He commands, "Up, one, Left, two," and the gunner, without changing sights, shifts his point of aim from the target to a point once its height above it and twice its width to the left. This method is practicable for all mass targets that have a dimension that may be readily gauged.

If the observer has an EE type field glass, the correction may be made as follows:

The observer looks through the glass, placing on the target the graduations of the inverted sight leaf scale which corresponds to the true range. He then notes what graduation is opposite the point where the shots strike. This graduation denotes the sight setting necessary to correct the vertical error. The number of yards the shots were short or over must also be estimated, and this amount added to or subtracted from the new sight setting to correct the horizontal error.

Example—see sketch: Range, 1000 yards. The burst of fire X, is observed at the 1100 yard mark on the inverted sight leaf scale of the field glass, and is estimated to be 100 yards short.

If the sight is set at 1100 yards, the vertical error will be corrected. Then 100 yards must be added to this sight setting to correct the horizontal error, making the new sight setting 1200 yards.

It is also observed that the burst of fire is 10 mils to the right. This lateral error can be corrected by taking 8

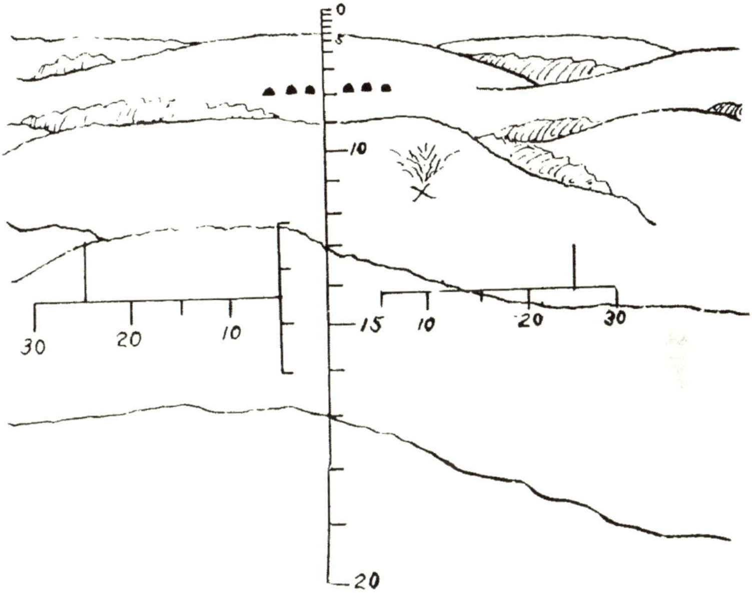

points left windage, since 1 and 1/6 mils equals 1 point of windage. If no field glass is available, and the strike of the shots is observed by the gunner, the correction may be made by moving the sight, *without moving the gun*, so that the line of sight is on the point where the shots were observed to strike. This movement of the sight should be made for either an error of elevation or one of deflection. The gun should then be shifted until the new line of sight strikes the target. This will correct the error of deflection, also the *vertical* error of range. In addition, the rear sight must be raised or lowered the amount of the estimated *horizontal* error of range, as described above.

When indirect fire is used, correction of fire, though attended by difficulties, is of prime importance, since the target is invisible to the gunner, and he is totally helpless to observe the effect of his fire except through the aid of the observer.

The correction of fire is rendered easier if the aiming point for indirect fire is as nearly as possible on a horizontal plane with the gun, for in this case the range on the sight leaf is approximately the true range to the target, and if the fire is, say 100 yards short, all that is necessary to correct this error is to raise the sight 100 yards. If however, the aiming point is so located that the sight setting varies widely from the true range to the target, such corrections will no longer give the desired result, as the following example will show. Suppose that the true range is 1000 yards, and, owing to the location of the aiming point, a sight setting of 2000 yards is necessary. If the shots are observed to strike 100 yards short, how much must the sight be raised to get hits? If the sight were raised 100 yards, that is, from 2000 to 2100, the shots would go too high, for the amount that we want to raise the sight is the distance between the graduation for 1000 and 1100 yards, and this is much less than the distance from 2000 to 2100.

The simplest way to determine the correction would be, if time permitted, to drop the sight leaf to 1000, and count the number of turns of the elevating screw head required to bring it to 1100, then reset the sight to 2000 and make the same number of turns. In action, however, time will not permit this method to be used, and the following approximate rule will be found accurate enough for all practical purposes.

Rule. To shift the fire 100 yards at any range, give the rear sight elevating screw head one quarter as many complete turns as there are hundred of yards in the true range.

In the example, the true range is 1000 yards, and the fire is 100 yards short. Therefore the rear sight should be raised ¼ of 10, or 2½ turns of the elevating screw head.

NIGHT FIRING:

Under modern conditions of warfare it will often be necessary to utilize machine guns at night to repel attack and cover certain sectors and restricted areas, (Bridge Heads, Fords, roads, etc.), over which it may be expected that the enemy will pass.

Accurate machine gun firing at night is practicable if the position is prepared before hand, and the proper methods of sighting are used. It is obviously impossible to illuminate the target, except with special apparatus, nor is it practicable to set a machine gun by day and expect it to stay put when firing from it at night. It will not retain its original aim beyond a burst or two, and if the aim be disturbed to reduce a jam, the gun cannot be reset, and is out of action.

This leaves the use of illuminated aiming points as the only solution of the difficulty. The best form of illuminated aiming point is the night firing box.

The night firing box consists of a box, (any kind will do), with one side covered with a semi-transparent covering, such as tracing cloth, oiled paper, ground celluloid, or the like. This transparent side has on it an aiming point, which will be either a black paster, or black lines. By day the sights are aligned on the aiming marks on the box, and at night these marks are made visible by illuminating the box inside with a flashlight or candle.

A little foresight and ingenuity will provide a box which will fold into a small space so as to be portable, and which should always be carried with the guns when in the field. Such a box is shown in the photographs. It is constructed so that it will easily collapse into a small flat object.

While the night firing box may be used at any distance from the gun, the one in the illustration is made to be used at a distance of 12½ yards from the gunner's eye. The vertical lines are 1 and 8/10 inches apart and the horizontal ones are 9/10 inches apart, so that at 12½ yards the vertical intervals represent 4 mils and the horizontal intervals represent 2 mils. The size of the box is immaterial. In the back is a hole just large enough for a pocket flashlight to be inserted.

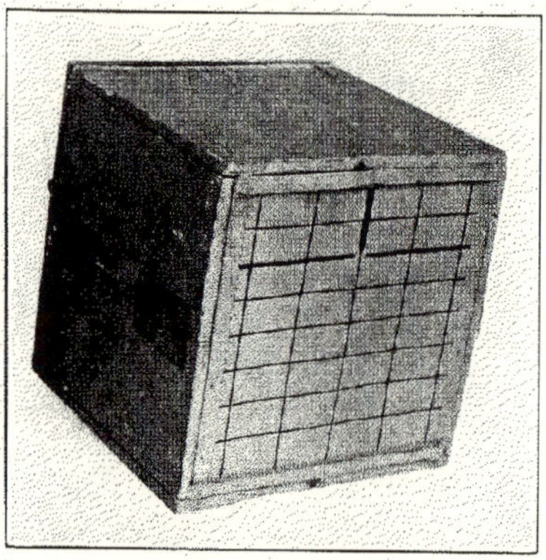

Collapsible night firing box

In the field it is possible to improvise a night firing box, using a pasteboard box or a tin can, with a piece of paper oiled with grease or neatsfoot oil.

Method of using the Night Firing Box.—Suppose that a bridge head, range, 1000 yards, is to be covered with machine gun fire during the night.

1. By daylight locate the gun, placing the muzzle on a notched stake, and with the correct range, 1000 yards, aim the gun at the bridge head.

2. Set out the night firing box 12½ yards from the gunner's eye, and approximately between the gun and the

target, but enough to one side of the line of fire to just avoid being hit. Mark the position by a stake.

3. Bring the sights to bear on the aiming point on the box, by the use of the elevation and windage screws, *taking care not to move the gun during this operation.* Note the necessary sight setting.

4. The gun and box may now be removed. At night the box is relocated by the stake and illuminated. The gun is placed with the muzzle on the notched stake, and with the sight setting determined in (3) is aimed at the mark on the box. Firing may now be commenced at any time, and if the aim of the gun becomes deranged, all that is necessary is to again aim at the mark on the box, and the target will be struck.

Note—If the gun is mounted on a tripod, the use of a notched stake for the muzzle is not necessary, the only thing required being to mark the position of the tripod legs and place the gun in the same position again in case it is moved.

A small can of luminous paint can always be utilized to good advantage in night firing for painting stakes, etc.

To conceal the distinctive flashes made by machine guns when they are fired at night, fire through wet sacks, blankets, etc.

To find how far in front of the guns a single aiming point must be placed to enable several guns to use it to cover a given front of the enemy's line:

It may so happen that only one night firing box is available for use by several guns. If all the guns aim at the same night firing box, their fire will cross, and the amount of the enemy's front that will be covered will depend on how far the box is from the guns. This is shown in the sketch. Suppose, for example, that a ford 1000 yards distant is to be covered by five guns. The ford has a frontage of 100 yards, and the 5 machine guns have a frontage of 25 yards. Where must the single night firing box be placed?

Procedure—Add the frontage of the guns to the frontage of the target and divide the range by the total.

2. Multiply the machine gun frontage by the result obtained in (1). This gives the distance that the night firing box must be placed in front of the gun position.

3. If circumstances permit it, this distance may be measured or paced and the box located accordingly, or it may be located as follows:

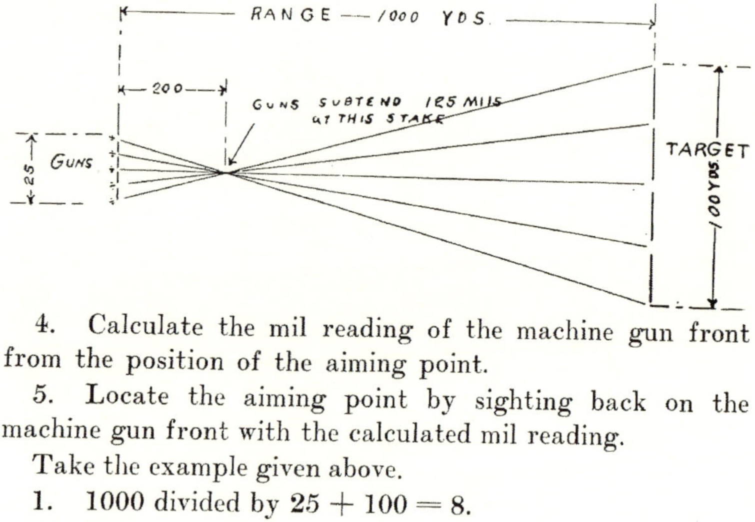

4. Calculate the mil reading of the machine gun front from the position of the aiming point.

5. Locate the aiming point by sighting back on the machine gun front with the calculated mil reading.

Take the example given above.

1. 1000 divided by 25 + 100 = 8.

2. 25 × 8 = 200.

3. The box must be located 200 yards in front of the gun position.

4. The mil reading of machine gun front from this point will be $M = \frac{1000W}{R}$, or $\frac{1000 \times 25}{200} = 125$ mils.

5. Go forward to point where machine gun front subtends 125 mils, and locate aiming point.

OVERHEAD FIRE.

As one of the important rôles of machine guns is to support the attack or cover the withdrawal, it sometimes becomes necessary for them to fire over the heads of their own advancing or withdrawing infantry. Since it is often impossible to obtain a flanking position, or any other position where the fire of the guns will not be masked, it becomes of prime importance to understand when it is practicable to begin firing over the heads of troops with safety, and also when to cease firing before their advance has proceeded far enough to bring them into the danger zone.

On account of the flat trajectory of the modern machine gun, it is impossible to use overhead fire when the guns, target, and friendly troops are all on the same horizontal plane, hence overhead fire can never be employed unless the guns or the target are on an elevation.

The "safety angle" is the angle between the target, the gun, and an imaginary danger line beyond which it is not safe for the attacking infantry to advance unless the fire be stopped, or elevated to establish a "curtain of fire" in the rear of the enemy's line to harass his retreat.

For our service ammunition there is not at hand any data upon which to base an assumption of the safety angle.

The German method described by Captain F. von Merkatz in his article on "A new Method of Machine Gun Fire" in the *Infantry Journal* is simply to elevate the point of aim 3 meters above the heads of their own troops. This seems to be principally for the benefit of the machine gunners, as it is not good for their morale to aim at their own men, even though theoretically the bullet is at a considerable height as it passes over the advancing line.

The British use two methods.

In the first method they consider the angle of safety to be 30′ for ranges up to 1000 yards, and 60′, (1°), for ranges from 1000 to 1500 yards. These angles are found by holding a 24-inch string to the eye and sighting past a card upon which are three lines. The first two lines are .21 inch apart, and the first and third lines are .42 inch apart. At 24 inches these distances correspond to 30′ and 1 degree, respectively.

In the second method they locate the danger point as follows:

1. Set the sights for the true range and aim at the target.

2. Without moving the gun, raise the sight 400 yards and locate a point by sighting.

3. This point is the danger point beyond which it is not practicable for troops to advance while overhead fire is being used.

This method is used for ranges up to 900 yards. For ranges beyond 900 and up to 1500 yards, the same procedure is followed, except that the sights are raised but 250 yards.

The use of the new type EE field glass would make this method particularly easy, as by means of the inverted sight leaf scale in the glass the squad leader can locate the danger point independently of the gun.

Owing to the extremely flat trajectory of our gun, the above angles of safety are not great enough for our use, and until some data on this subject is determined and promulgated to the service, it is suggested that greater angles than those mentioned above be used if the necessity for determining the danger point ever arises.

With the present lack of instructions on this point, each organization commander must use his discretion in the matter, but some definite way of determining the danger point is essential, otherwise the safety of the advancing line will be jeopardized.

PART III

MACHINE GUN TACTICS

The principles laid down in the following chapters are compiled from the writings of military observers and officers of the armies concerned in all the later wars, particular attention having been given to the principles developed by the present European war.

CHAPTER VIII

CHARACTERISTICS OF MACHINE GUNS

By the characteristics of any weapon, we mean the peculiarities appertaining to it. No matter how well trained the user is in the mechanical handling of his weapon, his efforts will fail if he does not use it correctly in a tactical sense. For this reason every machine gunner should understand the tactics of machine gun employment, and as the tactical employment of a weapon depends on its peculiarities, a study of the characteristics of the machine gun should form the ground-work for the study of its tactical employment.

Below are given the principal characteristics that govern the tactical employment of the machine gun. Some of the characteristics are unfavorable, but these are far outnumbered by the favorable ones.

First Characteristic: The Method of Mounting the Gun. Machine guns are always mounted on some sort of rest which wholly or in part relieves the gunner of the necessity of holding the piece. From this characteristic we get a reduction of the personal factor. With a machine gun properly mounted there is less work required of a tired and hungry man than would be the case were he handling a rifle which depended solely on the hold of the gunner for its mounting.

From the above considerations it follows that results obtained in time of peace with machine guns are comparable with these obtainable in war. This is not so with musketry fire. In an infantry company on the firing line there are many men handling many guns with many and different holds, and each man has nerves to be considered; while in the case of the machine gun there is one man, one barrel, one line of sight, one hold, and that hold one on which the nerves of the

Motorcycle with machine gun on detachable carriage

individual have no great effect. It follows that the cone of fire of a machine gun is much smaller and less variable than that produced by an equal volume of infantry fire.

This close grouping of shots produces a highly concentrated fire, which is particularly suited for surprise fire.

Machine gun fire is also most useful in the crisis of a fight, not only on account of its concentrated and accurate character but also because at this time when the riflemen have had their nerves racked to the utmost and have been fatigued to the point where accurate and rapid fire is an impossibility, the machine guns have not been affected and their superiority is more overwhelming.

The reduction of the personal factor in machine gun fire renders the observation of fire easier and more accurate than is the case with musketry. With machine guns if the strike of only a single shot it observed, it is certain that the others will be close to it. This is far from so with musketry fire, as the shot which is observed may be one of the many wild ones which are going nowhere near where the majority of the bullets are falling.

The fixed mount for machine guns also renders indirect fire possible. This is particularly useful for night work on outposts, etc. The gun may be laid by day so as to cover any threatened point and at night the gun can be fired with the certainty that the objective will be covered. In order to relay the gun when the vibration has disturbed the aim, the night firing box is used. This instrument is fully described in chapter V.

Second Characteristics: Rapid Production and Application of a large Volume of Accurate Fire. Fire can be produced rapidly, because the gun can be loaded and laid, and all that is necessary to produce a large volume of fire is to press the trigger.

Fire can be applied quickly because it is usually possible to get observation, and also because the correct point of aim can be picked up more quickly and with greater accuracy than with a number of rifles, for the machine gun officer has to make only one man understand his wishes instead of large numbers, as is the case with an infantry company. If there is any difficulty in indicating the point of aim the machine gun officer can sometimes lay the gun himself and thus indicate the proper target.

As regards rapidity of fire, machine guns shoot at the rate of from 400 to 800 shots per minute. This does not mean that the gunner actually fires this number of shots in a minute for the fire must frequently be interrupted in order to check the point of aim or reload the gun. The Colt and Maxim must be reloaded after 250 rounds, the Lewis after 47 rounds and the Benet after 30 rounds. This causes quite an interruption of fire. A further interruption is caused by the fact that the fire must frequently be stopped in order to check the aim. When machine guns hit, they hit hard, but when they miss they miss even harder, for they use up valuable ammunition to no purpose, and the supply of ammunition for machine guns is one of the greatest problems connected with their employment. For this reason it is desirable to check the aim after from 15 to 20 shots.

Firing in this manner, the actual number of aimed shots discharged in one minute varies from about 100 to 300. Experiments at the machine gun school at Harlingen in 1916 gave the following figures for the maximum rapidity of fire obtainable with trained gun crews, firing in bursts of about 20 shots.

With the Maxim, 360 aimed shots per minute.
With the Colt, 240 aimed shots per minute.
With the Benet, 180 aimed shots per minute.

With the Lewis, 141 aimed shots per minute.

It has been calculated that roughly one machine gun is equal to about 50 men if firing at rapid fire for one minute.

This characteristic of rapid production and application of a large volume of accurate fire makes machine guns particularly useful on outposts, as they can command any locality for any length of time, allow more men to rest, and can get into action more quickly and with probably greater effect than riflemen.

Third Characteristic: Narrow Front and Shallow Depth from which a Large Volume of Fire can be delivered. Only four or five square feet are required for the gun and its two men. But two riflemen could use their rifles properly in this space.

A machine gun can fire 200 aimed shots per minute. Two infantry-men firing at rapid fire can fire but 20 rounds per minute. The volume of fire that can be delivered by machine guns occupying a cramped space is therefore about ten times as great as that delivered by infantry occupying the same space. Hence the use of machine guns in cramped positions is obvious.

Fourth Characteristic: Invulnerability or Unassailability.—Invulnerability depends largely on invisibility. A gun in position with its men is a very small object, and is therefore easy to conceal, only fifteen inches of cover being required for guns on field mounts. Every effort should be made to conceal the guns, as if they are located they will be subjected to artillery fire. The vicinity of prominent objects should be avoided, as they may draw artillery fire. As few men as possible should be with the gun, and their disposition should be such as to simulate infantry whenever possible, thus making it difficult for the enemy to identify the machine guns.

Machine gun carriage detached from motorcycle

This should also be borne in mind when moving from position to position with light machine guns.

If the machine guns are located they are difficult to hit as they present such a small target. If they are brought under fire, they are difficult to put out of action, as the gun can be fired as long as one member of the crew remains. Casualties in the gun crews should be replaced by corresponding members from the reserve gun sections, who should be equally well trained.

If machine guns are located by artillery fire, the positions of the guns should be changed at once. When machine gun positions are being chosen alternative positions should also be selected, so that the guns can be moved if one position becomes untenable. The positions should be such that the change can be made under cover.

Fifth Characteristic: The all-around traverse. With a machine gun set up on a tripod, fire can be opened at once in any direction. This makes machine guns particularly suitable for use on a flank. Consider the case of a deployed company attacked from one flank. A sufficient change of front to deal with the flank attack would require some time, while the machine gun can open fire in the new direction instantly.

The possibility of having to fire to the flank should be considered when taking up positions with machine guns and when constructing trenches for them. It may sometimes be desirable to sacrifice cover for field of fire.

Sixth Characteristic: Mobility. As long as machine guns are carried on pack transport, they are more mobile than infantry and have about the same mobility as cavalry, as pack transport can go wherever horses can go. As soon as they are removed from pack transport, wagons or automobiles, they become less mobile than cavalry or infantry, as they have to be carried by hand, and men carrying them can only run a very short distance, and cannot advance by rushes on account of fatigue. Guns should be carried by pack or automobile as long as cover is available, then by hand.

Machine guns are classified according to mobility.

"The features forming the basis for the classification of machine guns may be stated as follows.

Light type, comparatively light weight, highly portable without tripod mount, air cooled, fed from a magazine.

Heavy type, relatively heavy with substantial tripod mount, effectively cooled for continuous firing and fed from a belt.

Each of these types has its special advantages and peculiar functions. The light machine rifle has superior portability and handiness, due principally to the absence of the tripod mount and cooling system, and to feeding from a magazine instead of from a belt. It can easily be carried by one man to considerable distances, and can be put into action very quickly. Such rifles are ideal weapons for supplementing the fire of riflemen, assisting them in gaining superiority of fire, and closely supporting them on all occasions, either in attack or defense. They can be used with the advance party or even with the point of an advance guard, and near the line of observation of outposts. In the defense they can be readily shifted to alternate positions to escape the effect of the enemy's artillery fire directed by aeroplanes. They are the first rifles to go forward in an attack, and the last to be withdrawn in a retirement; however, the construction of these rifles renders them unsuitable for long periods of continuous fire, and they should, therefore, be used in bursts rather than in protracted, uninterrupted fire. A typical example of fire by bursts is afforded by the fleeting opportunity of Aerial combat.

The heavy type of machine rifle is the automatic machine gun proper. While it can do most of the work of machine rifles of the light type, but with diminished celerity and greater effort, it has functions due to its stability and endurance which cannot be satisfactorily performed by any other type of rifle:

(a) To create one or more zones parallel to the general front of a defensive position which shall be so swept as to be untenable by the enemy:

(b) To guard the flanks of an attack against enfilade or envelopment by covering large areas near the flanks and rear with fire:

(c) Long range covering fire.

(d) Night firing to sweep positions, the direction and range of which are determined by day.

(e) Indirect fire.

(f) To provide sustained fire from defensive positions in fixed or semi-permanent works, an unlimited amount of ammunition being conveniently accessible." (Report of Machine Rifle Board, Nov. 10, 1916).

Seventh Characteristic: Liability to Accidental Cessation of Fire.—A machine gun is a gas engine in which the weight is exceedingly small in comparison with the horsepower. The Benet Machine Rifle as used in the U. S. Service has been calculated to be a gas engine of 59 horsepower. As the rifle weighs 29 pounds, this gives approximately ½ pound of metal per horsepower.

The weight per horsepower for aeroplane engines is 3.57 to 4.4 pounds; for high speed gasoline engine for racing boats, 15 to 25 pounds; and for steam turbine plants approximately 45 pounds. On account of the very high ratio of power to weight, the great velocity of the moving parts, and the necessity for the rifle to function hot as well as cold, mechanical troubles are bound at times to occur, and when the figures given above are considered, it will be realized that a great degree of mechanical excellence in machine guns has been attained to make them as reliable as they are.

Stoppages are classified as:

(1) Avoidable
(2) Unavoidable
 (a) temporary
 (b) prolonged

Avoidable stoppages are due to carelessness on the part of the gun crews, such as lack of oil, wrong adjustment, etc. They may be either temporary or prolonged. They should, of course, never occur, and with a well trained gun crew they will not occur.

Unavoidable stoppages may be due to faulty ammunition or to minor breakages, etc. They may be either temporary or prolonged.

Temporary stoppages are the kind usually encountered. They can be remedied by immediate action on the part of the gun crews.

Prolonged stoppages usually result from some serious damage to the gun, such as would occur if the gun were struck by a missile.

Eighth Characteristic: The noise of firing.—The sound of machine guns is an unmistakable sound—and has the serious disadvantage of instantly giving away the position of the guns. Machine guns have been found easier to locate when in the infantry firing line itself than when on the flanks of the line.

The noise of firing is not, however, always an adverse characteristic, for it has a great moral effect which encourages the men on whose side the machine guns are being used, and discourages the enemy.

Ninth Characteristic: Heat.—Machine guns become very hot after a few minutes of firing, and this heating makes the guns difficult to handle, interferes with sighting, owing to mirage, and is likely to give away the positions of the guns owing to smoke. It also causes rapid erosion of the barrels and quickly makes them inaccurate.

For this reason most machine guns embody some device for dissipating the heat. In the Benet and Colt this is accomplished by radiating flanges on the barrel. These flanges afford a large surface for radiation and for contact with the air, and help to cool the gun. When water is available, the guns may be cooled by direct application of water or wet sponges to the barrel. In addition, the barrels are made so that they can be quickly interchanged with spare barrels which are furnished with each gun.

The Lewis gun embodies a characteristic air cooling device consisting of an aluminum radiator surrounding the barrel, over which a current of air is drawn by each shot. The Maxim and Vickers guns are cooled by a water jacket surrounding the barrel. This cools the barrel effectively as long as the jacket is kept filled with water, but continuous firing gives rise to the generation of steam, which would interfere with the gunners aim and give away his position. For this reason a condensing device is also embodied.

Heating of machine guns not only interferes with the operation of the guns, but it also distorts the sheaf of fire and changes the range. In touching on this subject, Captain Friederick V. Merkatz, German Army, says, referring to the Maxim gun. "Should the barrel be not sufficiently covered with water, about two centimeters deep, parts of the barrel will be out of the water when it boils. This results, e. g., at a range of 1000 meters, in shortening the range as much as 200 meters. In addition the sheaf in this low shooting is materially enlarged and becomes so thin that the result obtained is out of all proportion to the amount of ammunition used."

CHAPTER IX

MACHINE GUNS WITH THE ADVANCE GUARD:

In the Russo-Japanese war the idea of the extreme value of the Machine Gun as a purely defensive weapon was thoroughly ingrained in both sides and as a consequence the capabilities of the weapon in this capacity were thoroughly developed. The Japanese used a single barrel air-cooled gun of the Hotchkiss type weighing 70 lbs. and mounted on a tripod weighing 40 lbs; while the Russian weapon was principally the Maxim. The extreme weight of these guns, the difficulty with which they were manoeuvered and the easy target which they presented to artillery and infantry fire during any attempt to advance them precluded their use as weapons of attack. In advance guard work they were usually carefully held with the main body and until the action had been well developed were not moved.

The European war found all the contending powers equipped partially at least, with a light gun capable of being advanced by hand with no more exposure than an Infantry skirmisher and with the greatest ease, consequently the Machine Gun has taken rapid strides as a weapon of attack, is not so carefully guarded as formerly, and has proved its efficiency in every engagement in which it has been employed.

The action of advance guard infantry is in general the same regardless of the size of the units engaged. Take a Brigade movement as perhaps most illustrative with a regiment of infantry forming the advance guard. The regiment then, will have four guns at its disposal, and perhaps more will be attached depending upon the character of the country, propinquity of the enemy, etc. In the event of a meeting engagement, more will doubtless be of great value. "Machine guns materially increase the effectiveness of an advance guard (F. S. R. 1914). They are especially useful in holding bridges, defiles and other key points until reënforcements can be brought up."

The desire of the advance guard commander will be to place his Machine Guns where they can most easily avail themselves of all opportunities to fulfill their functions and prove effective. This desired position varies with the formation assumed by the advance guard, due to the varying character of the country. If the country is close or broken the advance guard may be confined to the road, with patrols pushed out to the front and flanks. The regiment would perhaps put one battalion in support and two in reserve. The guns cannot be too much exposed and the fact that pack transport is used altogether in our regular service renders correct placing more difficult on account of having men and animals with their different rates of march in the same column. Now, if only four guns (the fifth being the reserve gun and as now issued without spare parts or ammunition mules for it) are available, one disposition would be to have two guns with the support, marching in rear and the remainder with the reserve. In very open country where the formation of the infantry of the advance guard will partake of the nature of a skirmish line the Machine Guns can be more usefully employed further to the front perhaps, say at the head of the support.

It must be remembered that machine guns are able to come into action and open a heavy and accurate fire in less time than it would take infantry of one-half its fire power to deploy and open a controlled fire from a position; it is therefore the machine guns of the advance guard which by opening fire should give the infantry time to deploy, find fire positions and develop an attack. Care must be taken not to keep the guns in action too long. After their function has been fulfilled they must be withdrawn and disposed according to their best tactical capabilities.

The company commander of the M. G. Co. must coöperate with the advance guard commander. After his guns have been withdrawn or have allowed the infantry to deploy, he must take his orders from the advance guard commander who should give him a free hand in every way, merely telling him his intentions and how he thinks the guns can be of greatest assistance to him. "All machine gun commanders must remember that local successes due to effective fire are absolutely worthless unless they further the object of the action, and moreover serve to uselessly expend an immense amount of ammunition." The guns having been withdrawn, can usually serve best on the flanks to bring a cross fire to bear upon the enemy, to fire on any bodies in close formations, led horses, artillery limbers, etc., that may present themselves. In taking up flank positions care must be used not to unnecessarily expose the guns by leaving them completely unguarded. The Commander should apply for at least a platoon of infantry to act as his guard and he should carefully reconnoiter all positions to be taken up. In doing this, he should send forward his scouts who should have had special training in the selection of positions, and if practicable the platoon leaders should precede the guns to their final accepted positions. Ranges should be at once determined, range charts constructed and given to the gunners, and all preparations thoroughly made before the guns reach the positions.

In the rencontre the value of the machine gun cannot be overrated. The advantage lies with the side which can first develop the strongest offensive. The principal characteristics of the rencontre are (I. D. R. 520). . . "An almost total absence of reconnaissance, necessity for rapid deployment, frequently under fire; and usually absence of trenches or other artificial cover. These conditions give further advantages to the offensive." All these characteristics favor the employment of machine guns. Reconnaissance lacking, the machine gun can develop, by fire, an enemy quicker than any other weapon; its ease of deployment and action have been touched upon, its deadliness at short ranges; and with a light field mount its comparative cover, render its presence a great advantage. During the first stages at least, of the rencontre, artillery support will be lacking and must be replaced by the use of machine guns for reducing points of strong resistance. Apropos of the use of the machine rifle, light type, with the advance guard, the Report of Machine Rifle board of which Brig. Gen. French was president submitted the following on November 10, 1916. " . . . . . They can be used with the advance party or even with the point of an advance guard, and near the line of observation of outposts. . . . . They are the first rifles to go forward in attack and the last to be withdrawn in the retirement; however, the construction of these rifles renders them unsuitable for long periods of continuous fire and they should therefore be used in bursts, rather than in protracted uninterrupted fire."

*Well trained Cavalry organizations can unpack and go into action in 25 seconds. What is believed to be the record time for Cavalry machine guns was made by the Machine Gun Troop, 9th Cavalry, at Camp Stotsenburg, P. I., Lieutenant Malcolm Wheeler-Nicholson, commanding, when before Colonel Dugan and the field officers of the regiment they went into action in twelve and two-fifths seconds, from the gallop.

*Army and Navy Journal.

MACHINE GUNS WITH THE REAR GUARD:

The rear guard is charged with the important duty of covering the retreat. . . . . It cannot, like the advance guard, count on the support of the main body (F. S. R. 1914). As shown by the European war and according to the best authorities, the only practical method of attaining good results from machine guns in combat is to place them all under one commander, a Brigade Machine Gun officer. Our present regimental distribution, with its consequent division of authority over the machine guns, renders this very difficult if not impossible. This fault shows up particularly in the formation of rear guards to cover a retreat.

The rear guard endeavors to fulfill its functions by causing the enemy to deploy at long ranges, by administering serious checks by ambush, etc., and by fighting a series of delaying actions. Care must be taken that the rear guard does not become so involved that withdrawal is difficult and too costly, for the rear guard can expect no assistance from the main body. The assignment of machine guns in this kind of an action must be governed by two considerations.

(1) The line of retreat must be protected, and (2), a strong rear guard must be formed. For the rear guard, extreme mobility is to be desired. The commander will need his machine guns to cover the withdrawal of his artillery, which will necessarily take place as soon as the hostile infantry has approached to effective rifle range, and the guns must be kept in action as long as the possibility of successful withdrawal is assured. The organizations most suitable for this work are Cavalry Machine Gun Troops. They will be called upon to make the greatest exertions, and, in face of an energetic pursuit, to suffer severe losses. They must be prepared to move from place to place with great rapidity and

for these purposes must retain their transportation for men and guns close to their firing lines.

Due to their compact beaten zones and the ease with which they can take advantage of small folds in the ground to establish a defensive position, machine guns are particularly suited for causing deployment at long ranges. They can quickly cease fire, pack and move out. Their power to seriously check an enemy caught emerging from a defile, or in thick or brush country to overwhelm him and withdraw before meeting an attack has been mentioned. It would seem then that with certain limitations machine guns, next to horse artillery, are the most important units of the rear guard and their presence will be decidedly felt. The retreat will be on a broad front and the rear guard must fall back on a broad front. Machine guns will be placed on the flanks where the terrain affords the enemy a likely chance of successful pursuit. Units less than platoons must not be separated and constant communication between flank companies must be carefully maintained. "The loss of men, horses and material must be unhesitatingly faced if there be any decided gain thereby" (F. S. R. 1914).

Machine guns of the rear guard are certain to come under artillery fire. Alternative positions are essential and it must be possible to reach them under cover and unobserved. Great care should be taken of the pack transportation for the loss of a gun mule will probably mean the eventual loss of a gun.

The possibilities of establishing successful ambushes are great and troops which can sustain successfully gradual losses of a considerable part of their personnel will become completely demoralized if the same loss is administered in a few minutes, as it will be when machine guns are used at close range.

*"In their retirements to successive positions machine guns must take care to fall back far enough to induce the enemy to reform column of march before coming under fire from the next position, for machine guns are very vulnerable to fire from skirmish lines with extended intervals." Every opportunity to employ cross fire or enfilade fire must be seized.

As one example of successful employment of machine guns with mounted troops in rear guard actions, General Alderson speaking of the South African War said: * "I had two Maxims with tripod mountings on pack saddles which belonged to the 1st Battalion Mounted Infantry. These guns had well trained mounted detachments, and a pushing officer, with a good eye for ground, in command. They were most useful, and more than once saved the flank of their unit from being turned by galloping up and coming into action on the flank of the out-flanking Boers. . . . . They can hold on to any such position as long as required to cover the advance or retirement of their infantry, and then easily catch them up or get into another position. In fact, if the detachment is mounted the value of the guns is more than doubled."

One or two companies should be kept in the reserve of the rear guard to be used in protecting the flanks as suggested by this example and must be under the immediate control of the rear guard commander.

In protecting the line of retreat the machine guns should be utilized to their fullest extent. "Points on the line of retreat whose possession is essential to the safety of the command, such as bridges, defiles, etc., must be secured in advance of the enemy." (F. S. R. 1914.)

Machine guns form a weapon especially designed for just these tasks if properly supported. The heavy type of guns will prove more advantageous for these duties, while the air-

*("Machine Gun Tactics.")—Applin.

cooled types will more satisfactorily perform the duties of the machine guns with the rear guard. The guns which are to be sent along the line of retreat to protect the bridges, etc., must be selected immediately by a staff officer and started to the designated positions. This is where the value of the Brigade Machine Gun officer is very evident. To an officer thoroughly conversant with machine guns, who has a thorough working knowledge of their tactical possibilities and also has a good eye for ground, the commander would only have to indicate the number of companies to be employed in this manner, his plan of action and line of retreat, with perhaps some of the most important points to be occupied. *"His instruction should be definite as to how long the points are to be held (until the infantry are past such and such a place, etc.); but should leave the details as to units of guns, opening fire, etc., to the machine gun commander who should be given a free hand in carrying out his instructions."

If guns are carefully placed it may be possible to surprise the enemy at bridge heads, in defiles leading to the line of retreat, or in narrow roads in brush country; but the guns, crew and animals must be carefully concealed. Opportunities such as this will not be infrequent for engaging pursuing cavalry during the early stages of the retirement and such an opportunity made most of has more effect than the most stubborn stand for it teaches caution better than anything else. Caution means delay and delay means a check in the pursuit such as to almost insure successful withdrawal. Broken units less than platoons should never be employed outside of the reserves. Frequently sections of a platoon will be able to employ cross fire and to support each other. Close coöperation is essential. Alternate positions easily accessible under cover must be selected and the position entrenched

* ("Machine Gun Tactics.")—Applin.

and prepared for stubborn defense. Usually the units will be held in positions in readiness near the position which they are to occupy and will prepare their positions before occupying them.

*"The Germans believe in Machine Guns thoroughly. Twenty-one German Machine Guns are known to have held a front of 1100 yards of trench for five months south of Soissons and to have repelled several attacks on this front." When machine guns are acting in these capacities alone, they must open fire upon suitable targets at long range. If they are supported by infantry they can reserve their fire until the enemy is caught at close range and overwhelmed. As soon as the machine gun company commander has established his company in position he should without any delay send his scouts or go himself to select fresh positions along the line of retreat. He should give the signal or establish the time for retirement. The fact that the guns are working in pairs and on a broad front should make the artillery's task of locating them not a particularly easy one. Under pressure a successive retirement of platoons under the fire of the retired platoon might be the most advantageous, but where close coöperation and support by infantry is possible, this will be found to be a slow and laborious process.

* "Machine gun and its development" by Neal Truslow, correspondent with the French Army at the Front. 1915.

CHAPTER X

Outposts and Combat:

The employment of machine guns with outposts varies widely according to:

(1) Types of guns available.
(2) Character of country.
(3) Nature of opposing forces.

With the outposts the guns may be used to sweep important approaches, hold bridges or defiles and to occupy certain important points which the enemy during his advance is likely to seize or to pass closely enough to afford a likely target.

The guns must not be used singly and just enough guns should be attached to carry out the above duties. A very few guns will usually suffice for these duties and the remainder of the number attached should be held in the reserve and with their pack transport ready to be moved to meet any attack.

Their functions with the outposts will be most important at night. Their great fire power renders it possible to considerably reduce the number of men necessary to hold important points; and by the use of the night firing box to fire with many times the accuracy of an equivalent number of riflemen.

The night firing boxes, of which a full description is given in Chapter VII should be placed in position after dark, if possible, and removed when it is light enough to shoot accurately without them. The guns should be placed to sweep with cross fire their particular sector, and it will usually be possible to accomplish this. In outposts then, the

Night firing box as used on outpost duty. The box is set 12 yards in front of the gun. It is placed by day for use at night.

chief problem becomes one of taking up positions, anticipating that these will be used to repel attacks made under cover of darkness. In doing this the following points must be borne in mind.

1. Artificial cover for the guns should be constructed, if it is possible to construct it unobserved.

2. The interval between the guns should be considerably lessened, for artillery fire at night is not particularly dangerous unless the position has been located by the enemy by day.

3. Field of fire and cover are points to be considered in selecting a position.

4. Flank defilade of at least five feet should be provided for each gun. In taking up these positions with the light type gun, it may frequently prove advantageous to advance all guns and ammunition by hand leaving the pack transport fully protected some distance to the rear. For the fixed positions of the outposts the heavy type gun is the better, while the guns with the reserve where mobility is desired might advantageously be of the lighter type.

The light type of gun unpacked may be used well forward, (depending upon the nature of the country), close up to the line of observation. It can easily be withdrawn, should the circumstances render this necessary, with no more noise or disturbance than that incident to the withdrawal of any infantry squad.

By using the night firing box accurate fire at night is possible without any apparatus for artificial illumination. With the night firing box, the steadiest and most accurate kind of fire is available. Even the poorest shots in the company under average day shooting conditions, will make excellent records by its use.

COMBAT:

1. *The offensive.* After the enemy has been developed it is expected that the infantry will deploy under the protection of artillery fire and will come as quickly as circumstances permit to effective small arms range. From this point on, after the completion of the artillery duel, the struggle becomes one for fire superiority.

After this point has been reached, the machine guns have their choice of two rôles. One to go forward with the attacking line; the other to select and occupy positions from which it may be possible to deliver an effective fire against the enemy up to the delivery of the assault. "Machine Guns are emergency weapons. They are best used when their fire is in the nature of a surprise to the enemy at the crises of combat. Their effective use will be for short periods of time—at most but a few minutes—until silenced by the enemy. When engaged they must be used to the limit of their effective capacity. On the offensive they find their use in assisting the attack to obtain fire superiority temporarily lost, and against lines of trenches which are to be assaulted. . . . ." (F. S. R. p. 172.)

The preparation by our artillery will have as one of its purposes the destruction of hostile machine guns as soon as they have revealed their positions. If these guns are not destroyed or their crews are not completely demoralized by the artillery, they will cause immense trouble to the attackers. If these hostile machine guns are located, the attacking machine guns may be able to dislodge them; however, in a contest of machine gun versus machine gun the advantage is going to lie with the squad having the best cover and this will ordinarily be the squad on the defensive.

The above is illustrated by the following from an account of a combat in the European War.

*"On the 9th of May the survivors of my company and of the adjoining company, about eighty men, arrived at 11 o'clock about 200 meters from the cemetery of Neuville-Saint-Veast. The cemetery being occupied, the field of battle seemed void of Germans. In the distance the batteries were fleeing. Two machine guns remained in the mill; this was the only resistance over an immense space, but it was sufficient. Impossible for my men to advance, we signal the fact with difficulty to the Artillery, which from this time on is under open field conditions; it opens fire a long time afterwards and mistakes its objective. Then before the eyes of our furious men, abandoned by all because they were too far to the front, the cemetery fills up with Germans. Four hours afterwards the 146th appears on the field and is mowed down by the machine guns; the next day the 229th succeeds it; the new repetition with a slight and extremely costly advance.

With these machine guns revealing themselves thus without our being able to forsee their emplacement, and taking up positions to stop our progress in a region no longer familiar to us, we must have means of suppressing them instantly. The field artillery is too far away; communication hardly exists after passing beyond the telephone lines. The position is of the greatest importance and merits study."

The machine gun squads should be drilled with the infantry and should participate in combat exercises with it. *"They should be accustomed to grasp the idea of the situation rapidly and to replace the fire of the attacking infantry either by taking a position in rear or on the flanks which will permit them to fire to the end of the action without being hindered by the movement to the front, or *by going squarely out in advance of the halted line.*

* (Capt. Laffargue—Inf. Journal Sept.-Oct., 1916)—"Impressions and Reflections of a Company Commander."

This last case should be thoroughly studied; the Germans have shown it to us and it is therefore possible; I know that it is very effective (25th of August at Crevic)."

The two types of guns, heavy and light, have their special functions in these rôles. The light guns, such as the Benet and Lewis with their portability and ease of operation with two men as a crew; with their capabilities for concealment, etc., are of the greatest use as weapons of offense with the firing lines.

The heavy type, the machine gun proper, finds its greatest use in firing over the heads of the advancing infantry. Its mounting makes it very well adapted for this. If the gun crew is well trained and the positions are well selected, these guns can continue their fire up till the moment of the assault without endangering friendly troops. They may be able to employ indirect fire in doing this and the problem of indirect fire under conditions of modern combat as shown in the European war, is deserving of great study and application. This is advocated by several British officers of note and instruments for determination of firing data have been issued in many instances to English machine gun companies. (See Chapter VII.)

Friendly troops can be fired over with entire safety by the heavy guns since the mounting, weight of the gun and tripod, and clamping devices permit of the certain control of the machine gun sheaf.* "In a flat country firing over the heads of the firing line is barred on account of the flatness of the trajectory. Theoretically, it would be practicable to fire over the heads of our troops without danger at the longer ranges on account of the height of the trajectory, but the line of sight would always be directed at our own troops or

* ("Machine Gun Fire"—Von Merkatz trans. Inf. Journal Sept.-Oct., 1916.)

pass slightly above them, so that such shooting is out of the question.

One requirement is to be able to distinguish friend and enemy from the position of the machine gun in order to prevent mistakes in observing the strike of bullets. In this kind of fire much will also depend upon the position held by the enemy. If he holds a commanding position so much the better. The steeper the slope in front of his position the longer the covering fire can be continued. The closer the attacking troops approach the enemy's lines, the more care, skill and knowledge will the officers and men of the machine gun companies need to show to prevent their own friends from coming into the machine gun sheaves."

A somewhat thorough knowledge of the ballistics and trajectory of the gun is necessary for this. (See Ordnance pamphlet No. 1923 for ballistic data for machine guns using .30 cal. government ammunition. The trajectory of these rifles is practically the same as that of the Springfield rifle, model 1903 and the data given in the above mentioned pamphlet is equally applicable.)

Firing through trees or brush over the heads of advancing troops must be avoided absolutely. At the Machine Gun School at Harlingen, Texas, in 1916-17 it was observed that even very small twigs and thick bunches of grass would deflect the bullets to a considerable and dangerous extent in this kind of firing. This class of firing has been used extensively in the European operation with great success.

*"The moral effect of firing over our own troops which it was feared would be demoralizing, was on the contrary reassuring and aroused the attacking spirit, as the men know that the enemy will be covered with heavy machine gun fire during their rushes.

*Machine Gun Tactics—Applin.

"Experiences in the field have demonstrated that the division of hostile fire caused by firing over our own troops was always marked; the losses were never as great as when the enemy fired on the machine gun and the firing line with the same sheaf." There are numerous examples of this covering fire in both the Russo-Japanese and the European wars.

The idea of employing machine guns in the firing lines is comparatively new. Before the European war its use in the firing line was considered dangerous for the following reasons.

(1) It attracted artillery fire and so endangered the firing line which it was endeavoring to assist.

(2) It was too conspicuous. Its gun crews were too much exposed and very vulnerable.

Machine guns squad advancing "as skirmishers." The gun is carried loaded, and can be put into action almost instantly. These few men can carry the gun and 1,000 rounds of ammunition almost indefinitely.

(3) Its ammunition supply, the amount which it was thought possible to advance by hand, was considered too variable a proposition to be seriously considered.

With Machine guns of the air-cooled type, the gun squads should be trained to deploy, advance by thin lines, etc., using exactly the same tactics as infantry units. If these tactics are followed out, it will seldom be possible for artillery observers to locate them until they have gone into action, and then it will be too late, for guns used in the firing line will be kept out of action until they are able to engage the

Machine gun squad advancing in squad column

enemy at close range. They must be willing to lie quiet and if possible await the moment of the assault, or the time when fire superiority must be regained.

With the air-cooled gun, the crew necessary to operate it occupies only the frontage of two riflemen. Any little defile or fold in the ground is large enough for it. If the squad leader has a good eye for selecting positions the gun can be advanced across ordinary terrain without any more exposure than is incident to the advance of two infantry soldiers. The light weight of the air-cooled gun renders it almost as easily portable as a rifle of the ordinary type.

With suitable positions and good background it does not expose its infantry unnecessarily to artillery fire nor does it prematurely disclose its position. The gun itself presents a small target to artillery and it is difficult for them to adjust upon it, even when it is stationary and well located.* "The 75 is employed to destroy the machine guns. Unfortunately on account of the dispersion, it does not perfectly fulfill its rôle; its shots often fall to one side and a great number of them are often necessary to find exactly the small space that holds the machine gun."

"The Germans use machine guns in attack boldly and cleverly. They push them up close to hostile trenches, and in this way sometimes prepare the way for the Infantry attack. In the attack on our positions these guns have sometimes been brought up by snipers before the infantry advances. Cases have even occurred in which they have brought their guns into farm buildings 10 or 15 yards in front of our trenches, and have rendered the latter untenable."

**"Machine guns are used in large numbers against one or both flanks of the portion of the position which it is intended to attack. They usually cross their fire, which makes them difficult to locate from the portion of the trenches opposite them. An attack has even been carried out solely with machine guns. The trench was engaged from a flank by six or seven guns, while other machine guns succeeded in working round and enfilading the position."

In using guns with the attacking line the Germans have placed them behind the second or third wave. They have

*The Attack in the Present War—Laffargue—Inf. Journal Sept.-Oct., 1916.

**Machine Gun Training—Capt E. J. Solano.

done this with the idea of protecting them until they could get within short range, about 400 yards, before their advance and that of the infantry with them should be stopped. As soon as the advance was stopped the machine guns crawling out well in front of the halted line have opened fire, becoming "the teeth of the attack."

*"The exact range is usually obtained by opening bursts as soon as a suitable fire position has been occupied, after which the Germans satisfy themselves by preventing the defenders, as far as possible, from showing above the parapet, thus enabling their own troops to approach in security. The closer they can approach a trench, the more oblique becomes their fire. The duration and volume of the fire depend upon the ground over which the advance of their own infantry is to be made, but they are careful to husband their ammunition, as the ammunition supply is the chief difficulty of these guns.

When the advance of their own infantry has passed the machine guns, the Germans try to place the latter in positions where they can assail the enemy as he retires from his trenches or alternatively, in the event of a counter-attack to open fire in such a way as to allow their own infantry to withdraw."

**"The Machine Gun is an element of attack and the most terrible arm of close fighting. However, it is employed in the attack only to man the positions taken or to support the infantry elements from a distance. This is nonsense; to give it such a rôle one could never have trembled with rage and impotence at a few paces from the enemy whom he could not get at.

The machine gun should be pushed as far as possible in front of the halted line of fire. If it remains behind or abreast of the fighting line, its field of fire is generally blocked

*Machine Gun Training—Capt. E. J. Solano.

**The Attack in the Present War—Laffargue—Inf. Journal Sept.-Oct., 1916.

or masked by the slightest movement. In advance of the line it will enable the infantry line to advance for some time under cover of its fire. It can move forward; its crew of a few men can creep along the smallest pathway and a shell hole is sufficient for its shelter; in the skirmish chain a whole ditch is necessary. Will it lack ammunition, having only the boxes which the crew carries sometimes incomplete? No, for it has only to fire on rare occasions, for example at the moment of the assault."

This method of employment both in offense and defense seems to have met with success in several specific instances during the early part of the European war while open fighting was still in progress.

The German gunners always took advantage of the great surprise effect and grasped every opportunity to take advantage of the confusion created among the enemy by this fire. Each member of the German Company is an expert in handling the gun and the Germans, who seem to have made a greater study of machine guns and their fire effect than any other nation consider it to be the most effective way to discharge bullets; accordingly they have organized their machine guns as a separate arm, "under trained permanent gunners, as they evidently consider that none but specialists can attain the necessary efficiency."

The only safeguard against the overwhelming losses of surprise with machine guns is very careful reconnaissance. The ease with which machine guns, particularly those on the defensive and working over known terrain, can be concealed adds to the possibility of catching troops in close formation and attaining surprise.

At Gallipoli, from an apparently deserted beach, machine guns cleverly placed caught the British under a cross fire and caused great losses. A machine gun can be hidden where no other weapon larger than the shoulder rifle can.

After the attack has been pressed home, if it is successful, all available machine guns should be advanced to the position to secure it against counterattack or to complete the demoralization of the enemy by joining in the pursuit. In this latter rôle Cavalry Machine gun troops will play the greater part. If they coöperate carefully with the pursuing Cavalry and the horse artillery it will be almost out of the question for a demoralized enemy to reform.

MACHINE GUNS ON THE DEFENSIVE:

The characteristics of the Machine Gun render it the weapon of all weapons for defensive purposes. The heavy gun with its compact sheaf here outshines the lighter air-cooled type and shows up to greater advantage. The first consideration in the employment of the guns in the defense is the choice of positions and the method of occupying them.

Machine guns are readily destroyed by artillery and concealment must be afforded, so if time permits cover must be constructed. To select a position in ordinary terrain that will permit a combination of these requirements to be fulfilled will usually be the result of a compromise in the mind of the person selecting it and will mainly depend upon the following considerations:

(1) Good field of fire or one easily prepared within the time available.

(2) Securely held flanks or flanks resting upon some natural feature which will assist materially in rendering it impregnable.

(3) Sufficient frontage to allow all guns to be placed at wide intervals and if possible not on the same line.

(4) Good communications throughout the position.

(5) Demolition of all buildings, telephone poles, etc., in rear which would be liable to act as artillery markers for the enemy.

In addition, the nature of the foreground should be considered as to the ease with which obstacles can be constructed and placed in effective positions. Unless the defense is to be a purely passive one and the machine guns are not to go forward throughout the action, much care will generally be necessary in placing these obstacles so as not to impede the movements of the defenders.

The character of the soil, whether or not favorable to ricochets, should not be overlooked although in the compromise between the requirements of a defensive position and the tactical requirements, this point is usually the first one sacrificed.

Alternative positions for each pair of guns should be at once selected and prepared but not occupied by any of the personnel. Covered approaches to it must be available and if the natural features do not afford such cover, it must be constructed. The guns should be assigned to their positions separately but should never be placed outside of mutual supporting distance so that a temporary stoppage of fire or the destruction of a gun will not endanger the position.

In assigning sectors and in clearing the field of fire, if such action is necessary, the sectors must be arranged so that the fire of each gun crosses that of one of the others; or so that there is a possibility of successfully striking the advancing enemy in flank. Each gun must be carefully and cleverly concealed and not mounted in its position until its presence is necessary. The artillery reconnaissance will ordinarily be carefully made and in their preparation for the attack, if any inkling of the location of the guns is suspected by them, the guns and their positions will be quickly destroyed.

A standard form of emplacement particularly as to the shape and size of loopholes must be avoided. All loopholes should be closed by a sand bag or by some other suitable arrangement when not in use. Too much care in concealing the guns is impossible. Head cover is very necessary where time for its construction is available. (For some types of entrenchment see Chapter XII.)

*Some of the things to be avoided are indicated in the following:

"The Machine Gun emplacements are recognized in the continuous trenches by the *low horizontal loopholes* much larger than ordinary loopholes. They are generally quite easily recognized; occasionally the machine guns are in a little separate work which is quite characteristic."

In order to obtain the maximum effect with machine guns, the fire must be reserved until the latest possible moment when a really good target has been presented. It is absolutely inexcusable for a machine gun to open fire from a concealed position until it can get the greatest possible fire effect from its weapon. "To ensure this it is essential that the target be big and vulnerable and at *close range.*"

In the Russo-Japanese war instances were cited by the observers showing clear examples of the right way and the wrong way for guns in such positions, to come into action. **" . . . . On January 27, 1905, a Japanese company attacked Sha-Shan. Four Russian Machine guns opened fire at about 1100 yards upon an extended firing line, without causing any serious loss or affecting its advance.

*The Attack in the Present War—Laffargue—Inf. Journal Sept.-Oct., 1916.

**Applin—Machine Gun Tactics.

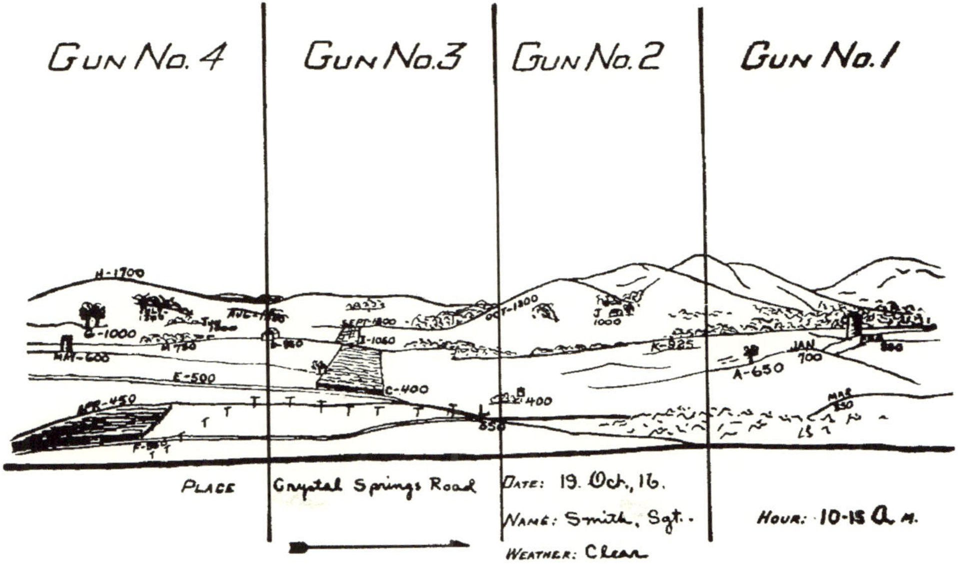

Range chart for defensive position. The sector to be covered by each gun is indicated

On the other hand on March 1 the Japanese had approached to within 200 or 300 yards of the Russian position at Wang-Chia-Wo-Pang and were beginning the final assault. Two Russian Machine Guns suddenly came into action, and the Japanese assault was repulsed with heavy loss due to its annihilating fire."

This has been borne out in the European War conclusively. When time permits, its is made a requirement with machine guns, to place them in positions where they can catch the attackers under cross fire (this is usually on the line of obstacles), or to place them so as to enfilade the attacking lines, and to sweep all dead spaces.

This applies only after the direction of attack has been carefully ascertained. (Under conditions of trench warfare

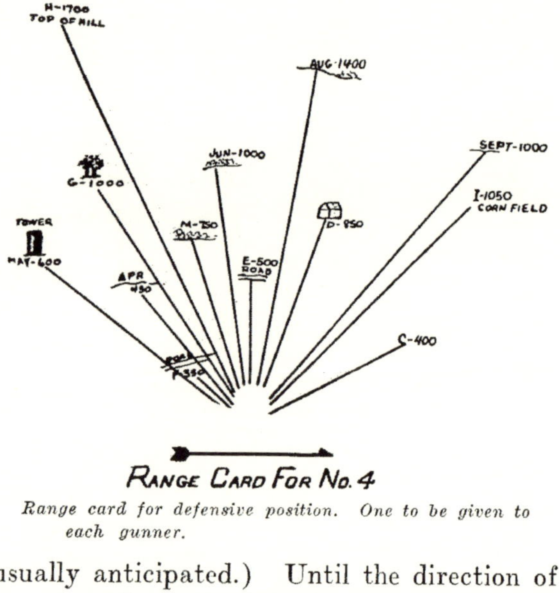

RANGE CARD FOR NO. 4

Range card for defensive position. One to be given to each gunner.

this is usually anticipated.) Until the direction of attack has been accurately determined, it is dangerous to place too many machine guns on the front line of defense and they will be found of great value in the reserve, to prevent unexpected turning movements, repelling assaults, delivering counter attacks and reënforcing any weak points which may develop in the line of defense.

Their presence in the front line will render it possible to reduce the number of riflemen employed there and so to economize in troops without weakening the position. After the position has been selected and during its preparation, range charts should be constructed showing ranges to all prominent objects and all dead spaces. These dead spaces should be identified by a letter or a short name so that fire data can be readily communicated to adjoining units or to those units which can fire effectively against these spaces.

The simplest range cards possible should be given to each gunner and he should be carefully oriented as to his position.

A "well" placed some distance to the front of an established defensive position. It is carefully concealed and fire is reserved until the moment of the attack, in order to catch the attacking force in the flank. Concealment has been removed in this case, to permit the construction of the trench to be shown.

The accompanying card and chart are types which have been worked out and found successful in the European War.

During the early part of the European war while open fighting was still in progress, the German Machine Guns, handled by well trained squads were frequently placed well to the front of the general line and were given instructions to hold their fire at all hazards until they had an opportunity to catch the attacking hostile line in flank. This they were able to do in quite a number of cases in the majority of which the German machine gun emplacements were missed in the preliminary artillery preparation. One account of such a combat follows:

*"At one point German Infantry to the number of about two hundred had been placed on the crest facing across the bare level plateau, while in front of them some two hundred and fifty paces distant was a pine wood through which the French were advancing. The Germans had evidently had no time to entrench but had quickly lain down in skirmish order in the outer edge of a potato field; each soldier had then pushed up in front of him, as protection a little heap of potatoes and loose earth. A hundred paces to the right of this German skirmish line, two mitrailleuses had been skillfully thrust forward some fifty yards in advance, and concealed in small trenches hurriedly dug. They could thus fire across the front of their own infantry and take in flank any French who advanced. This action was one of a series which had taken place along this line of hills. The German flanks were not unprotected but owing to the fact that the country was much broken and obscured by woods, such a force would be partly hidden from its neighbors to the right and left and largely independent in repelling any attack made against it.

* Notebook of an Attaché—The Battle of the Marne—Wood.

View of the "well" trench as it would appear in action, from a distance of twenty feet

A body of French infantry three to four hundred strong had advanced to the edge of the wood facing the Germans and had there taken up a skirmish position. The opposing bodies had then fired at each other a collective total of about 25,000 rounds across a perfectly flat field. . . . . When we arrived upon the scene the wounded had nearly all been removed, but the dead were still untouched, and we were able to see that, as a result of this fusilade of about 25,000 rounds, only three Germans and six Frenchmen had been killed outright.

After this rifle contest, the French had made a bayonet charge across the open. The Germans had fired until the French had advanced about half way and had hit a score, after which they temporarily ceased firing and the French then promptly 'charged home.' The two German mitrailleuses were unperceived by the advancing French, and as the French passed them in flank, the mitrailleuses opened fire; at the same moment the Germans suddenly fired a scattering volley. Attacked in front and on the flank, every Frenchman but one was hit, and sixty dead still lay in a row across the field as if cut down by a mowing machine."

Incidents of the successful use of machine guns in the defense of positions are very common in all of the modern wars. It seems to have been well understood from the first that such was their true rôle up to the time when the machine gun was replaced by the automatic machine rifle.

All experience has gone to show that the machine gun handled by well trained gunners and properly placed tactically is of the utmost value, while if in the hands of poorly trained gunners or handled wrong tactically it becomes an absolute menace to its own personnel as well as to any neighboring friendly forces.

CHAPTER XI

TRENCH WARFARE

The peculiar conditions and situations brought out in the present European war have been the result of having two opposing forces of about equal effectiveness, with impregnable flanks, fighting in a restricted area. These situations were formerly considered impossible on account of the increasing mobility of all arms of the service and accordingly all training and development for this kind of warfare is based on actual experience obtained in the trenches after the outbreak of war.

The nature of the trench combats is somewhat similar to that of siege warfare; the gun positions being more or less hastily chosen and afterwards carefully prepared.

In this kind of warfare the machine gun is brought squarely under the fire of its arch enemy, artillery, and must carefully prepare its positions to be able to exist for any length of time. Late reports coming from Europe indicate that in order to resist successfully the heavy artillery fire which always precedes an attack, undisturbed cover of at least 20 feet in that portion in which the personnel is quartered is considered necessary. This great depth of trench has placed such a premium on speed in handling and portability of the machine gun that the day of the heavy type in Europe seems to be surely over. During these bombardments all guns and men are withdrawn from the firing positions to the deep shelters, for nothing could live during the delivery of this "drum fire" as it is called.

Guards equipped with periscopes are left at the positions and report the progress of the advance. As soon as the "drum fire" is lifted, the whole result depends upon the outcome of a race between the attacking and the defending force. If the machine guns, grenadiers, etc., are placed in their positions in time to catch the attackers under an effective fire, the attack usually fails. If on the other hand the attacking force reaches the parapet in advance of the machine guns, the position is generally lost. There has been developed a

Rear view of emplacement for Maxim

considerable amount of additional material, hand grenades, periscopes with sighting attachment, etc., which experience has shown is necessary to successfully meet the requirements of this kind of warfare. Shields are used to some extent with machine guns and have been found to be effective. Personal armor aside from the steel helmet has so far proved unsuccessful.

The greatest of care must be taken in the selection of positions for the guns and in occupying them in such a way as not to attract attention. The site selected must be located

Same Maxim emplacement as it appears from a position 25 feet to the front

so as to give the machine guns either flank or cross fire. Frontal fire has been found of value with machine guns only in very exceptional circumstances. The emplacement must be carefully concealed, not be in front or in the vicinity of conspicuous objects which would serve the enemy as ranging points, and every effort must be made to make the position tenable, and if time permits as comfortable as possible for the crew of the gun; while at the same time the position must ensure if possible, that the attacking force will be surprised by fire.

All emplacements must be provided with overhead cover, splinter proofs, wire protection against hand grenades, and should be so built that its crew can enter or leave it easily

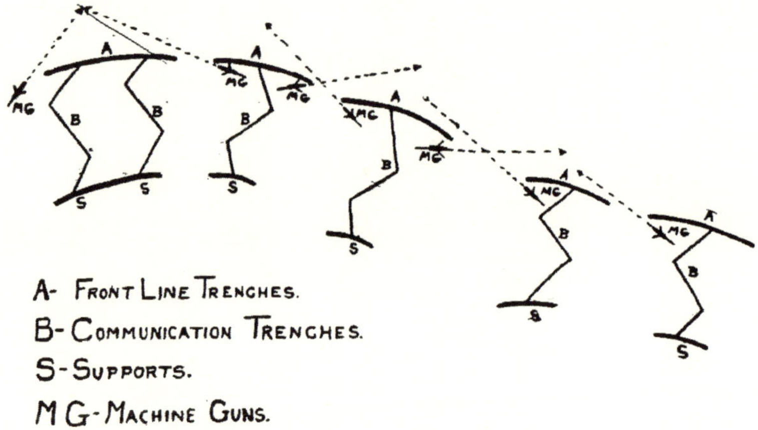

Plan of part of an entrenched position, showing suitable positions for machine guns

and unobserved. Enough room must be provided in the firing trenches for spare ammunition, hand grenades, etc. Inasmuch as the guns are exposed to fire from the front if placed in enfilading positions, they must be protected from such fire by a good thick traverse. Entrenching tools must be retained close at hand and extra sandbags are found to be of great value in hastily repairing damages.

Some positions in which machine guns have been found effective in the European War are roughly illustrated in the accompanying sketches.

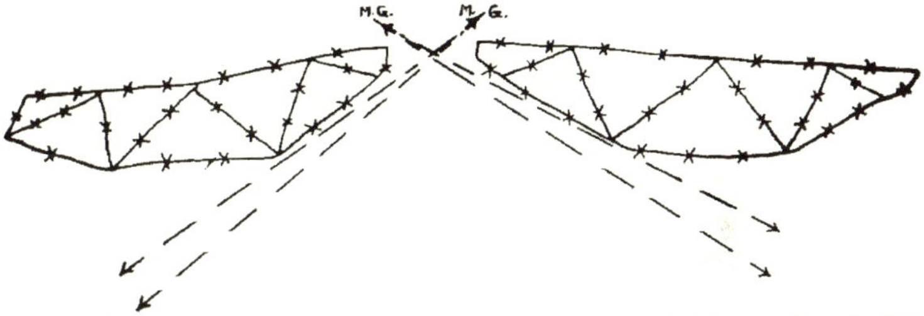

Portion of wire entanglements, showing gaps purposely left to lure the attackers into the fire of machine guns.

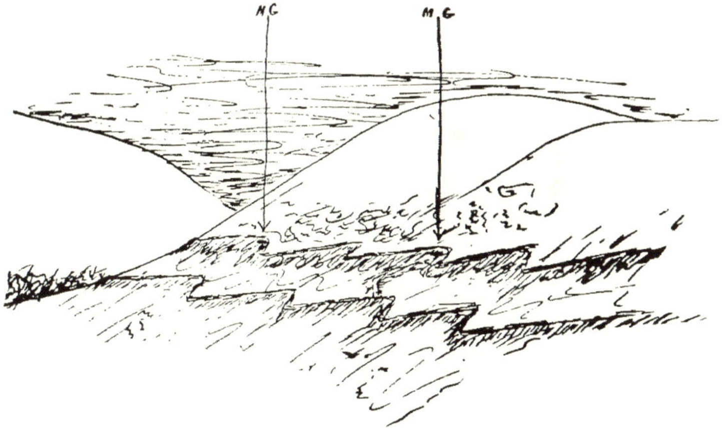

Showing how machine guns may be used in second line trenches when elevation of the ground is sufficient to permit such a disposition.

Gaps have frequently been left purposely in the lines of wire entanglements and a pair of machine guns mounted in these gaps. When the attackers have crowded towards these apparent lines of least resistance they have had to face the murderous fire of machine guns.

Well out in front of the general line of trenches so placed as to be able to obtain a flanking fire against an advancing enemy. These emplacements are usually "wells" of the type shown in Chapter XI and are connected up by smaller branch trenches which are either subterranean or so narrow and well concealed as to be practically invisible.

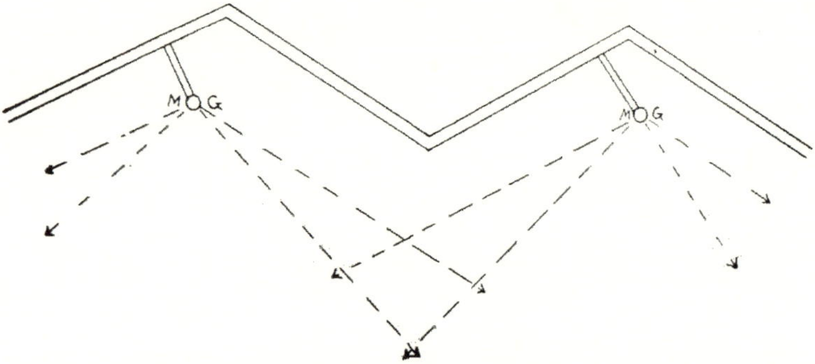

Machine gun "wells" in front of a defensive position

In attempting to provide concealment for these "wells" it is usually impossible to construct head cover as that would render the "well" conspicuous and it is liable to be silhouetted against the main trench; however, care must be taken to insure concealment from aerial reconnaissance. This is done by using branches, grass, etc.

For each pair of guns inside the trenches there are provided alternative positions which are carefully prepared beforehand but not occupied. The guns set up outside the trenches are not moved nor do they fire under any consideration until they have an opportunity of fulfilling their functions of catching the enemy in flank. This they have succeeded in doing in a number of cases.

Many ruses for the successful use of machine guns have been employed by both combatants in the European War. A description of one follows: In reëntrant angles one branch of the trench was continued but the addition was cleverly concealed by a strip of painted canvas to resemble the rest of the trench. Behind this canvas machine guns were placed to command the interior of the entire trench. As soon as

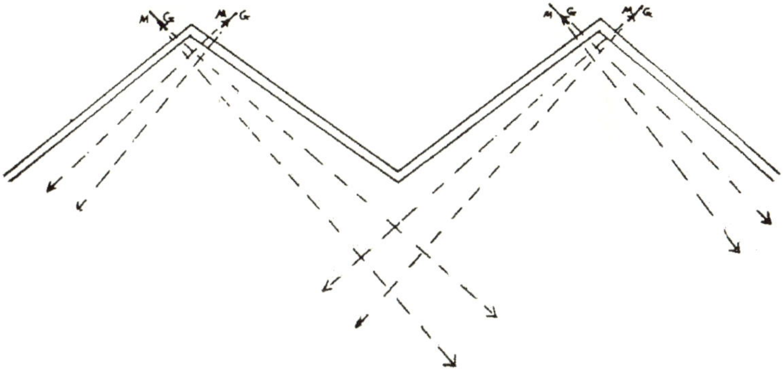

Re-entrant angles in a line of entrenchments are usually defended by machine guns, as from such positions they can obtain a flanking fire as shown.

the trench was occupied by the attackers, a burst of fire from these concealed machine guns practically annihilated the enemy and made the trench untenable.

Dummy works and dummy guns have not proved of a great deal of value during the present war probably due to the very congested area in rear of the lines. "When it is desired to conceal the existence of a whole position, dummy works must not be made at all."

At night much concealment has been afforded machine guns by the artificial illumining apparatus. However, under ordinary conditions this apparatus is handled by other troops and its methods do not concern machine gun companies.

*"(1) No fixed apparatus can be set in the field for the reliable lighting of a foreground at night which will survive a prolonged bombardment and bad weather, or admit of renewal or attention in the close presence of the enemy. This does not apply to electric searchlights employed in carefully chosen and protected sites, in connection with harbor and fortress defenses, nor to flares or bonfires set for the purpose of providing a single illumination for a given purpose, such as a signal or landmark.

(2) For the temporary illumination of foreground during an engagement with the enemy a good method is to employ hand and rifle 'illumination grenades' which ignite on impact with the ground, upon the same principle as in the detonation of the percussion hand and rifle grenades. By this means temporary illumination can be concentrated at the exact points at which it is most required. In Siege warfare or trench fighting at close quarters there is special use for a combined incendiary and illuminating bomb or grenade fired from a trench mortar."

For this trench warfare it is essential that members of the company be trained in using grenades, periscopes, and night firing devices. Night firing with machine guns from a position entrenched (the emplacement being a part of a generally prepared position) has not been found very successful. The position is quickly disclosed and spotted from the characteristic jet of fire which emanates from the gun at night, and as a result is subsequently destroyed.

In building trenches for the first line the size should be just sufficient to afford cover to the crew of the gun; this will prevent the more curious men of the company or those idle for the time being from crowding in and so increasing the casualties if the enemy scores an unlucky hit. Before the position is completed, if it can be done without taking

*Field Entrenchments—Edited By E. J. Solano, p. 217.

A machine gun center of resistance in the second line of defense

chances of exposing the position, have all the guns placed, and let your gunners check up on the field of fire and their sectors therein. Much dead space can sometimes be corrected by a few minutes work spent in slightly altering a loophole. It has been found very difficult to control and

Light type machine gun emplacement. Only sufficient room is necessary for three men.

prevent the destruction under cover of darkness of obstacles placed more than thirty of forty yards in front of trenches employing frontal fire.

A short description of a German position follows:

**"The German defenses comprised the first and second positions. The first position consisted of three or four lines of trenches, covering 400-500 yards in depth, separated from each other by barbed wire entanglements. The second position consisted of a single trench with here and there a support trench. This second position and its barbed wire entangle-

** French Offensive in Champaigne—Capt. X—May, 1916.

ments were built almost entirely on the reverse slope of the hill, where it was most difficult for the artillery to get the range. In addition, wherever the ground permitted, little organized centers of resistance and nests of concealed machine guns had been formed. The German troops had orders to cling to these to the last in case the other trenches were overwhelmed."

CHAPTER XII

ENTRENCHMENTS FOR MACHINE GUNS

The General Principles to be followed in entrenching machine gun units are the same as those for infantry as regards the object of the fortifications, the method of construction, amount of cover necessary, the importance of concealment, the use of dummy works, etc., and all these subjects may be found in any of the standard manuals on field fortifications.

For this reason, no attempt is here made to go into details regarding these things, and the scope of this chapter is consequently limited principally to a discussion of certain points which are of particular interest to a machine gun commander.

The types of entrenchments to be used will differ radically in different cases, depending on whether or not they are constructed in a defensive position, which has not yet come under the fire of the enemy, or whether they are hastily constructed during a lull in the engagement.

Entrenchment During the Attack.—In the attack, the primary object is to dislodge the enemy, and this requires continued progress to the front. For this reason, the use of entrenchments during an attack will be comparatively rare, and will, in general, be limited to those cases where the attack has been halted, and where it is desired to hold the ground already gained until circumstances permit a further advance. In this case, a very few minutes of work with the entrenching tool will suffice to provide a measure of protection.

A type of hasty entrenchment which may be used in the case mentioned above is shown in the photograph. A shallow pit is scooped out for the gun and its crew, and the earth thus obtained is thrown up to the front and sides to form a parapet. As the light type guns on field mounts are very close to the ground, and as the Vickers gun mounted in the lowest position requires only 15 inches of cover, it will be seen that a fair amount of protection can be obtained with comparatively little work.

Hasty entrenchment

The chief disadvantage of entrenching in offensive operations is that once the attacking force is fairly entrenched, the sense of security obtained may discourage further advance on the part of the attackers. This tendency should be recognized and carefully guarded against.

The Use of Natural Features.—It goes without saying that every advantage should be taken by the attacking force of any natural cover that may be available. Either raised or sunken features may form valuable cover for machine guns. Some of the raised features that may be used are road or railroad embankments, stone walls or fences, etc.

Sunken features are road or railroad cuts, ditches, natural folds in the ground, etc.

Machine Guns Covering the Attack.—A distinct use for machine guns in the attack is found in the case where circumstances permit them to be posted in the rear of the attacking line in a position which has sufficient elevation to allow them to fire over the heads of their own forces and thus keep down the enemy's fire. In this case, the machine guns should be intrenched if time and circumstances permit it.

In such cases the necessity for alternative gun positions should be considered, for if a machine gun is located by the enemy it will become the objective of artillery fire which will render the position untenable. If an alternative position has been prepared in advance, it can then be occupied, leaving an abandoned position for the enemy's artillery to destroy.

Entrenchments in Defensive Operations.—These will, in general, be much more elaborate and effective than the hasty entrenchments described above, as their construction is usually accomplished before coming under fire, and with less limitations as to time.

Whether machine gun entrenchments made under these circumstances are separate emplacements or part of a general line of trenches, their construction will be similar, and will, in general, include the following features:

Cover.—Sufficient depth of excavation and height of parapet must be provided to allow the gun crew to stand or sit while serving the gun.

Head Cover must be provided if time permits. This means that the gun will have to fire through a notch or loophole.

Overhead Cover, as protection against shrapnel or plunging fire, will be provided if possible.

Loading Pits or sheltered positions near the gun with sufficient space for reloading the belts or feed strips should be included.

A Lookout Station, or point where one of the members of the gun detachment can observe the field of fire, will often be necessary.

Concealment of the Position is most important. If a part of a line of trenches can be recognized as a machine gun position, it will receive special attention from the enemy's artillery. Care should be taken to avoid the vicinity of buildings or other prominent marks which might draw artillery fire. The position should be located so as to avoid being silhouetted against the skyline. In case loopholes are used, it is of the greatest importance that they should be closed

This illustration shows many points to be avoided. Nearly everything is wrong. The trench is located against the skyline, and no background has been placed behind the trench to prevent the heads of the occupants from being silhouetted against the sky. No attempt has been made to close the loophole with sand-bags when not in use. The partly demolished building in the rear would furnish an excellent marker for the enemy's artillery.

by some such obstruction as sandbags, branches, or a bunch of grass, stuffed into the opening when it is not in use, to avoid its being recognized for what it is.

A machine gun emplacement where these requirements have been disregarded is shown in the photograph. Such a position would not long survive the enemy's artillery.

The accompanying sketch shows one type of entrenchment for machine guns of the light type. Hasty cover is first constructed, and as time allows it, this is deepened and elaborated as shown in the second and third steps. In order to economize time, the emplacement is made no larger than necessary. No parapet is used, all the dirt being thrown behind as a support for the overhead cover, and as a background to render the occupants of the trench invisible.

An example of a trench improvised by Captain Carpenter, 2nd Indiana Inf., especially to accommodate a Lewis gun used to cover the approaches of a bridge is shown in the photographs.

Trench for Lewis gun in process of construction

EMPLACEMENT FOR BENET OR LEWIS.

FIRST STEP. HASTY COVER

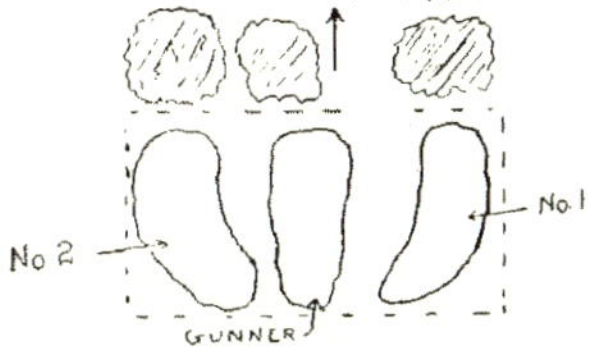

Gunner and members of crew dig shallow pits - one foot deep where emplacement is to go.

Time required - five minutes.

SECOND STEP. EMPLACEMENT.

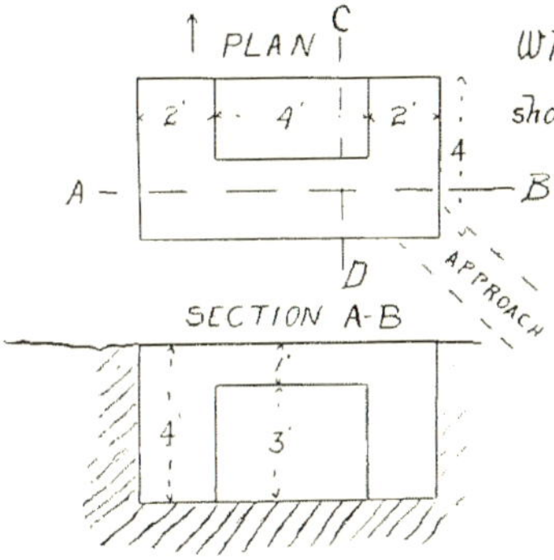

When time permits - hasty cover should be converted to form shown.

Dirt is thrown to the rear.

Time required - one hour.

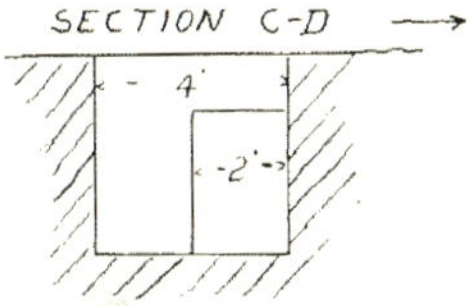

THIRD STEP. OVERHEAD COVER.

Additional time required - one hour

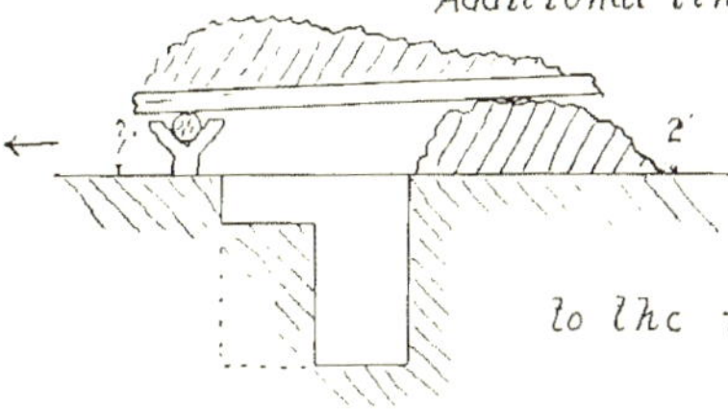

Note that dirt is thrown to the rear.

Lewis trench after completion

The form of emplacement that will be used in any case will depend not only on the ingenuity of the officer in charge, but also on the circumstances of the action, and for that reason no set form of trench is recommended here, as none can be adopted as a standard to meet all conditions.

Spotted something that isn't quite right?

Let us know and we'll fix it for future reprints.
And as a thank you for your diligence, we'll
send you a free Sothis book of your choice.

publisher@sothis.press

This page intentionally left blank

Made in United States
Orlando, FL
04 November 2026

90218335R00141